Chemical Pretreatment
of Nuclear Waste
for Disposal

Chemical Pretreatment
of Nuclear Waste
for Disposal

Edited by

Wallace W. Schulz and
E. Philip Horwitz

Chemistry Division
Argonne National Laboratory
Argonne, Illinois

Springer Science+Business Media, LLC

Library of Congress Cataloging-in-Publication Data

On file

Proceedings of an American Chemical Society symposium on Chemical Pretreatment of
Nuclear Waste for Disposal, held August 1992, in Washington, D.C.

ISBN 978-1-4613-6076-6   ISBN 978-1-4615-2526-4 (eBook)
DOI 10.1007/978-1-4615-2526-4

© 1994 Springer Science+Business Media New York
Originally published by Plenum Press New York in 1994
Softcover reprint of the hardcover 1st edition 1994

PREFACE

Chemical pretreatment of nuclear wastes refers to the sequence of separations processes used to partition such wastes into a small volume of high-level waste for deep geologic disposal and a larger volume of low-level waste for disposal in a near-surface facility. Pretreatment of nuclear wastes now stored at several U.S. Department of Energy sites ranges from simple solid-liquid separations to more complex chemical steps, such as dissolution of sludges and removal of selected radionuclides, e.g., ^{90}Sr, ^{99}Tc, ^{137}Cs, and TRU (transuranium) elements. The driving force for development of chemical pretreatment processes for nuclear wastes is the economic advantage of waste minimization as reflected in lower costs for near-surface disposal compared to the high cost of disposing of wastes in a deep geologic repository. This latter theme is expertly and authoritatively discussed in the introductory paper by J. and L. Bell.

Seven papers in this volume describe several separations processes developed or being developed to pretreat the large volume of nuclear wastes stored at the US DOE Hanford and Savannah River sites. These papers include descriptions of the type and amount of important nuclear wastes stored at the Hanford and Savannah River sites as well as presently envisioned strategies for their treatment and final disposal. A paper by Strachan et al. discusses chemical and radiolytic mechanisms for the formation and release of potentially explosive hydrogen gas in Tank 241-SY-101 at the Hanford site.

Three other papers are included that illustrate both the great diversity of waste pretreatment technology and the types of waste that are candidates for pretreatment. For example, the paper by Johnson et al. discusses technology that might be applied in the future to wastes generated from reprocessing of integrated reactor fuel. Microbiological and/or photochemical techniques that may be applicable to pretreatment of certain liquid nuclear wastes are advanced in the paper by Francis and his colleagues.

The final paper, by Hemmings, et al., presents some innovative technology for pretreatment of mixed wastes.

We believe the 11 papers in this book provide, to expert and neophyte alike, an excellent overview of the current status and multidirections of nuclear waste pretreatment research and development activities in the United States.

W.W. Schulz
E.P. Horwitz

CONTENTS

Separations Technology: The Key to Radioactive Waste Minimization 1
   J.T. Bell and L.H. Bell

Chemical Pretreatment of Savannah River Site Nuclear Waste for Disposal 17
   D.T. Hobbs and D.D. Walker

Disposal of Hanford Site Tank Waste 25
   M.J. Kupfer

Process Chemistry for the Pretreatment of Hanford Tank Wastes 39
   G.J. Lumetta, J.L. Swanson and S.A. Barker

Removal of Actinides from Hanford Site Wastes Using an Extraction
      Chromatographic Resin 51
   G.S. Barney and R.G. Cowan

Chemical Mechanisms for Gas Generation in Tank 241-SY-101 71
   D.M. Strachan, L.R. Pederson, S.A. Bryan, E.C. Ashby, C.L. Liotta,
      E.K. Barefield, H.M. Neumann, F. Doctorovitch, A. Konda, K. Zhang,
      D. Meisel, C.D. Jonah and M.C. Sauer, Jr.

Combined TRUEX-SREX Extraction/Recovery Process 81
   E.P. Horwitz, M.L. Dietz, H. Diamond, R.D. Rogers and R.A. Leonard

Noble Metal Fission Products as Catalysts for Hydrogen Evolution from
      Formic Acid Used in Nuclear Waste Treatment 101
   R.B. King, A.D. King, Jr., N.K. Bhattacharyya, C.M. King and L.F. Landon

Microbiological Treatment of Radioactive Wastes 115
   A.J. Francis

Treatment of High-Level Wastes from the IFR Fuel Cycle 133
   T.R. Johnson, M.A. Lewis, A.E. Newman and J.J. Laidler

Soil*EX^{SM}- An Innovative Process for Treatment of Hazardous
      and Radioactive Mixed Waste 145
   G.C. Gilles, M. Husain, R. Hemmings and R. Neuman

Clean Option: An Alternative Strategy for Hanford Tank Waste Remediation;
   Detailed Description of First Example Flowsheet 155
      J.L. Swanson

Index 211

SEPARATIONS TECHNOLOGY:
THE KEY TO RADIOACTIVE WASTE MINIMIZATION*

Jimmy T. Bell[1] and Luci H. Bell[2]

[1]Chemical Technology Division
[2]Office of Environmental Compliance and Documentation
Oak Ridge National Laboratory
P. O. Box 2008
Oak Ridge, TN 37831

INTRODUCTION

Waste and waste management problems are as old as the human race. For purposes of this discussion, wastes can be grouped and identified according to three chemical categories: organic, inorganic, and radioactive. Throughout history, humans have disposed of organic wastes by piling them on the earth or by burial. Until this century, most of the organic waste that was disposed to the environment was manageable by natural biodegradation processes. Even though kinetically slow, these natural processes seemed to satisfactorily manage organic wastes until the middle of the twentieth century. During the post-World War II years, the combined effects of increasing population and accelerating industrial production created organic wastes in quantities that began to overwhelm the natural biodegradation processes. New chemical technologies began to produce very stable chemical products specifically designed to be used once and then discarded as waste. Many of those stable materials, such as plastics and transformer oils, are very resistant to biodegradation.

Historically, humanity's primary means for managing inorganic waste has been to discharge it into the earth's crust. These processes sometimes included a direct discharge to the earth where rainfall later diluted the inorganics to low concentrations. Sometimes the discharge was into streams which rapidly dispersed the materials. About the same time that we began to realize our organic waste disposal rate was exceeding the biodegradation rate, we also realized that, in some locations, discarded inorganic materials were accumulating to hazardous levels in our environment. Some easy-to-recognize examples of these inorganic wastes are the cadmium and chromium from electroplating processes and silver and cyanide from photoprocessing operations.

*The authors wish to express appreciation for the time and effort contributed by many reviewers across the DOE complex. Their comments and constructive criticism have improved the quality of the resulting paper. The views and interpretations presented here are those of the authors and do not necessarily reflect the views of the Department of Energy, Oak Ridge National Laboratory or Martin Marietta Energy Systems, Inc.

Chemical Pretreatment of Nuclear Waste for Disposal, Edited by
W.W. Schulz and E.P. Horwitz, Plenum Press, New York, 1995

1

Beginning in the 1940s, we also began to accumulate a type of hazardous waste that is totally resistant to biodegradation, and which we realized could not safely be discharged into the earth's crust. This category of hazardous materials includes all the radioactive by-products and wastes from nuclear power production, national defense activities, and nuclear medicine.

Management practices for hazardous waste in the United States during the first 30 years of the war period seem almost primitive by present standards. There was a general attitude that waste was something that could be thrown away and forgotten. Radioactive wastes, too, were put out of sight and out of mind as quickly and cheaply as possible, while our efforts and resources were directed to more urgent production needs.

A growing environmental consciousness that began in the 1960s and 1970s led to the enactment of environmentally protective legislation such as the Clean Water Act, the Resource Conservation and Recovery Act (RCRA), and the Clean Air Act. Legal standards and requirements in the '90s have become increasingly stringent in mandating a cultural change in the production and management of waste by-products, including hazardous organic, inorganic, and radioactive chemicals.

The "throw-it-away" waste management practices in the post-war United States, and in other rapidly developing countries then and now, placed priority on economic development and emergency needs. Postwar economics drove U.S. industries and government to quickly dispose of wastes to the environment. Items that could not be directly disposed of were stored in fields, tanks, etc., to be managed later or to become waste legacies. Often, the needs for rapid production of defense materials and low-cost production of consumer goods significantly affected waste management decisions and practices in U.S. defense and industrial activities during the postwar and cold-war periods.

As we approach the end of the twentieth century, we are becoming increasingly aware of the many types of hazardous waste streams pouring into our environment. These wastes are produced by research and production processes, by natural and man-made disasters, and by discard of residual input materials. In recent years, while there is a worldwide cultural movement toward conservation and environmental restoration, a new parameter is interfering with intelligent waste management decisions and practices. Environmental activist groups apply pressure to industry and government agencies to quickly remove waste from particular areas, without adequate planning and consideration of what should be done with the waste once it is removed. Public perception and political interests, fueled by media attention, strongly support (and often demand) a "fast relief" approach to a legacy of waste insults to the environment. Such demands are coercing the United States to act once more with a "fix-it-all-quick" attitude, rather than develop solid, scientifically based waste-management strategies. In some ways, this demand for "fast relief" or "quick fixes" from some environmental groups reflects an attitude similar to that which, in past years, was responsible for creating our most serious problems, i.e., the attitude that the problems should be eliminated as quickly as possible. Instead, we need to ensure that the problems are eliminated as safely and as effectively as practical.

There is a better way. As scientists we must provide leadership in conserving world resources and preventing pollution through better planning, more efficient processing, and improved technologies. We must also resist the "quick fix" mentality and assist our national leaders in establishing sound waste management strategies that will both conserve our economic resources and protect the environment. Hazardous waste materials are sometimes pure or nearly pure chemicals; however, most hazardous wastes contain large amounts of innocuous and frequently useful components mixed with lesser amounts of one or more hazardous components. One of the best approaches to hazardous waste minimization and efficient waste management may be a simple 1-2-3 combination: (1) separate, (2) recycle, and (3) minimize disposal. By separating the lesser quantities of hazardous components from the larger quantities of innocuous components, and by recycling as many of the useful components as possible, we can achieve the third objective of minimizing the waste requiring disposal. This concept is the same used in all residential waste recycling programs, where the household segregates waste according to glass, metal, plastic, paper, and rapidly decomposing items. The idea is so simple that it appears obvious. One might expect that everyone would be willing to devote the time and effort needed to separate and plan for properly managing most wastes. But the long-established American custom of quickly abandoning waste has been very difficult to change. Americans are committed to "progress" and expansion; the idea that waste disposal requires and deserves significant investment of money and technological expertise has not been widely recognized.

The basic concepts of separation, recycling, and minimizing disposal are equally applicable to hazardous waste management in a production scenario and to remediation efforts directed toward the legacy of abandoned waste. Minimization of the hazardous waste should be the primary objective of separations technology. Chemical technologies suitable for waste separation include (1) solvent extraction, (2) ion-exchange chromatography, (3) precipitation, (4) dissolution, (5) solvent leaching, (6) filtration and membrane processes, and (7) combustion and degradation processes. These can be utilized alone or in combinations, depending upon waste disposal requirements, the waste stream, time constraints, and available resources. The application of any of these technologies to waste management and waste separations will require adaptation of specific processing details, but each of these technologies is very advanced in concept and in practice.

There are also many effective approaches to minimizing waste production. One way is to stop using processes that, while making a product, also produce an unacceptable or undesirable by-product waste form. This almost obvious approach has been adopted at several DOE sites, primarily with respect to production of nuclear defense materials. Another effective means of minimization is to change the chemical processing to reuse or recycle specific components, thereby reducing process wastes. This too is being applied at several DOE sites. A third approach is the primary use of separations to minimize hazardous waste disposal. This has not been widely applied, but effective use of separations technologies can be done at many government sites, including the DOE sites.

The "separate/recycle/minimize disposal" approach is particularly well suited to the legacy wastes such as the DOE radioactive wastes stored in tanks at various sites. The sheer magnitude of this "legacy waste" problem, the complex nature of the materials, and the budgetary constraints imposed by a reduction in economy make a compelling case for the application of separations technology to this reduction effort. For example, rather than solidifying the tank wastes and trying to find a way to "bury-it-all-again," we encourage DOE to chemically separate the waste components into a smaller volume that will require expensive disposal as hazardous waste and larger volumes of less hazardous wastes (requiring less-expensive disposal) as well as usable or recyclable components.

The remainder of this discussion will address such a separations-based waste minimization approach to the management of the DOE legacy radioactive tank wastes.

DOE LEGACY TANK WASTES

Management of DOE wastes has, in the past, varied considerably, depending on the specific DOE site and the particular objectives at the site. Several of the DOE sites have produced large quantities of aqueous radioactive wastes that were buried in underground tanks of 37.8 to 3780 x 10^3 m^3 (10^4 to 10^6 gal) capacity. These wastes are complex mixtures in which the bulk of the chemicals are nonhazardous; some components may even have commercial value. The various DOE tank wastes stored at four DOE sites are discussed here.

Oak Ridge National Laboratory (ORNL) Site

In the early days of the Atomic Energy Commission (AEC), ORNL developed the major processes for separation of plutonium from irradiated uranium fuels. This early development of plutonium separation processes included several separations technologies such as precipitation, filtration, solvent extraction, and ion exchange used to isolate the plutonium from fission products.[1] A host of other separations technologies were developed later and used in producing radiochemical isotopes for medical and research purposes.

Wastes from the early nuclear research and development studies at ORNL were stored in large gunnite tanks that were buried underground in a centralized zone of the present ORNL site.[2] The structural integrity of these tanks, built in the 1940s and 1950s, became a concern in the 1970s. The ORNL and AEC conducted studies on the use of hydrofracture for waste disposal and, as a result, decided to pump this waste into the underlying geological structures, safely isolating it between layers of shale at 1000 to 1200 ft underground.[3] The radioactive liquid waste was mixed with grouting materials and pumped under high pressure through sleeved wells into the rock formations under the Oak Ridge Reservation. About one-half of the gunnite tank wastes were disposed of in this way, until some strontium-90 contamination was observed in the associated test wells. The use of hydrofracture disposal was then discontinued in 1984, and most of the remaining gunnite tank wastes were subsequently transferred to the newly built Melton Valley Storage Tanks (MVSTs).

Since the completion of the eight 189 m^3 (50,000 gal) MVSTs at ORNL in the 1970s, all liquid radioactive wastes generated at the Laboratory have been stored in those tanks. Since 1985, the waste levels in the tanks have been near capacity, and several avoidance campaigns have been carried out to provide tank space for newly generated wastes.[4] In these campaigns, supernatant liquid was removed from the tanks [a total of ~189 m^3 (50,000 gal) in each campaign] and converted into grouted waste forms for on-site storage.

Although grouting produces a stabilized waste suitable for long-term above-ground storage, we tend to overlook the fact that this technology actually <u>increases</u> the volume of radioactive waste that must then be given long-term environmentally protective storage. It is neither economical nor desirable from a long-term waste management perspective to continue in this mode.

The Hanford Site

The AEC established the Hanford site in 1943 to produce and recover plutonium. Various reactors and fuel reprocessing plants were operated there until 1985 for this purpose. Less significant objectives were recovery of cesium-137, strontium-90, and neptunium-237 from the spent fuels. Production of weapons-grade plutonium from uranium requires low fuel burnup (less than 1%) for high plutonium-239 content in the plutonium product. This means that the production of a ton of plutonium required several orders of magnitude greater tons of uranium fuel to be irradiated and reprocessed.

The reprocessing technology for the spent fuel at Hanford followed the same pattern as the developments at Oak Ridge, and the associated wastes were made basic with NaOH and stored in underground tanks. The early tanks for this waste storage were built in the 1943—1964 time frame and were made of concrete with a single carbon-steel liner. These tanks have been designated as the single-shell tanks (SSTs), and most have volumes of about 3780 m^3 (10^6 gal). The tanks were designed for lifetimes of <40 years, and some of the SSTs are known to have leaked. At this time, these 149 tanks have had the liquids mostly removed so that the remaining contents are solids or sludges which do not leak into their environment.

The tank construction at Hanford changed from SSTs to longer-lasting double-shell tanks (DSTs) in the 1970s. The DSTs are also 3780-m^3 (10^6-gal) tanks. Each has two steel liners with an outside concrete support structure. A primary advantage of the double lining is that the space between the liners can be monitored for leakage. Currently, all of the DSTs are structurally sound, and the hazardous waste contents are mixtures of solids and liquids. A common description of the DST contents is a supernatant liquid over a sludge, where the sludge may occupy as much as 50% of the volume. There are 28 DSTs buried at Hanford, and they are filled to near capacity. The total tank waste volume at Hanford is ~9 x 10^4 m^3 (2.4 x 10^7 gal) in the DSTs and 1.4 x 10^5 m^3 (3.7 x 10^7 gal) in the SSTs.

The Idaho Site

The Idaho site has processed spent Navy nuclear fuels over the lifetime of the nuclear Navy program until 1992, when all naval fuel processing ceased. This site converted the fuel reprocessing waste into a calcined product. Calcination reduced the waste volume and produced an easy-to-store solid material. However, the calcining process was discontinued by DOE order in 1992, leaving about 189 m^3 (5 x 10^5 gal) of acid reprocessing waste. Many DOE contractor staff believe that the proper management of this acid tank waste would be for DOE to permit restart of the calcination process, allowing the Idaho site to completely calcine all of the reprocessing waste to a solid form and store it with the formerly calcined waste.[5] The Idaho site also has ~5680 m³ (1.5 x 10⁶ gal) of high sodium nitrate acidic waste from decontamination efforts, and that site is planning a technology to manage this waste with the sodium-depleted portion to be calcined. The retention of the acid wastes in the tanks may be the most hazardous option for managing this waste.

The Savannah River Site

This site has produced plutonium and tritium by irradiating uranium and lithium targets, respectively, in production reactors. Processing of the targets to recover the plutonium or the tritium has produced large volumes of liquid waste. Waste management at the Savannah River site has focused on producing salt cake from much of the liquid waste, and this is currently stored in vaults. However, there are remaining about 42 tanks of 3780 m³ (10⁶ gal) each that contain alkaline waste with supernate, salt cake, and sludge materials. Also, the Savannah River site is planning to begin, in this decade, vitrification of the transuranic (TRU) waste into ~0.6-m³ logs for subsequent storage in the national long-term high-level wastes repository.[6] The vitrification unit is almost completed, and the feed preparation for vitrification is being formulated.

Comparison of the DOE Site Tank Wastes

The tank wastes at all four sites are of great concern to DOE. However, Assistant Secretary of Energy Duffy in 1991 identified the tank wastes at the Hanford site as presenting the greatest potential hazard to the environment; current Assistant Secretary of Energy Grumbly reiterated this concern in 1993. A major portion of that hazard potential is the 177 radioactive waste tanks, and especially the 149 SSTs that have high potential for leakage. An average composition of the Hanford SST tank waste is shown in Table 1. Analysis of these data reveals that the dominant chemicals in the Hanford tanks are sodium nitrate and sodium nitrite, and the major secondary components are various aluminum compounds.

The composition of the ORNL tank wastes is similar to that of the Hanford tanks (Table 2). The dominant tank waste chemical at ORNL is also sodium nitrate. The primary difference in waste composition at the two sites is that the major secondary chemical at ORNL is calcium oxide, rather than the aluminum compounds found at Hanford. Characterization of the Idaho and the Savannah River tank waste shows great variance from that of the Hanford tanks. The similar characteristics of the Hanford and ORNL tank wastes, and the DOE concern for hazard potential of the Hanford tanks, provide justification for a pilot project to develop applications of chemical separation technologies to minimize DOE tank wastes.

Description of the Hanford and ORNL Tank Wastes

Based on the data in Tables 1 and 2, the Hanford and ORNL tank wastes can be described as chemical wastes containing primarily sodium nitrate contaminated with RCRA materials (primarily Cr and Hg) and having low levels of radioactivity. The 177 Hanford waste tanks contain 2.3 x 10⁸ liters of waste that are contaminated with an average radioactivity of 1 Ci/liter. This can be compared to the typical radioactivity of dissolved spent fuel from Purex processing, which is ~500 Ci/liter.[7] A similar analysis of the data for the ORNL MVST wastes indicates <0.1 Ci/liter radioactivity.

Table 1. Bulk components of Hanford SST wastes.

Component	Total quantity estimates[a]			
	Moles	Grams	Isotopes	Curies[b]
Al	9.05×10^7	2.44×10^9		
Am		$\sim 10^4$	^{241}Am	3.6×10^4
Ba	4×10^{6b}			
Bi	1.25×10^6	2.61×10^8		
Ca	3.25×10^6	1.30×10^8		
Cr	2.21×10^6	1.15×10^8		
Cs		5.6×10^{5b}	^{137}Cs	9.5×10^6
Fe	1.12×10^7	6.27×10^8		
Hg	4.49×10^3	9.00×10^5		
K	9×10^{6b}			
Lanthanides	$6 \times 10^{5b,c}$		^{151}Sm	$\sim 6 \times 10^5$
Mn	2.18×10^6	1.20×10^8		
Na	2.25×10^9	5.17×10^{10}		
Ni	3.04×10^6	1.78×10^8	^{63}Ni	$\sim 3 \times 10^5$
Np		4.7×10^{4b}		
Pd		6.7×10^{6b}		
Pu		3.8×10^{5b}	^{239,240}Pu	2.7×10^4
Si	7.93×10^6	2.22×10^8		
Sr	4.11×10^5	3.60×10^7	^{90}Sr	4.5×10^7
Tc		$\sim 5 \times 10^5$	^{99}Tc	$\sim 9 \times 10^3$
Th		1.3×10^{7b}	^{232}Th	1.4
U		1.4×10^{9b}	^{238}U	4.7×10^2
Zr	2.70×10^6	2.46×10^8		
Cl^-	1.13×10^6	4.00×10^7		
CO_3^{-2}	2.68×10^7	1.61×10^9		
F^-	4.24×10^7	8.05×10^8		
$Fe(CN)_6^{2-}$	1.52×10^6	3.22×10^8		
H_2O	2.49×10^9	4.48×10^{10}		
I^-		5.6×10^{5b}	^{129}I	2.4×10^1
NO_3^-	1.56×10^9	9.67×10^{10}		
NO_2^-	1.04×10^8	4.80×10^9		
OH^-	5.38×10^8	9.15×10^9		
PO_4^{-3}	9.20×10^7	8.74×10^9		
SO_4^{-2}	1.72×10^7	1.65×10^9		

[a]From values in DOE-EIS-0113 [Ref. 11] unless indicated otherwise.
[b]Separate estimate.
[c]Midpoint between 1.7×10^6 in DOE-EIS-0113 [Ref. 11] and 2.0×10^5 in separate estimated quantities of bulk components.

Table 2. Comparison of Hanford and ORNL tank waste constituents (water-free basis).

Type	Classification	Hanford tanks			ORNL Melton Valley tanks	
		Liquid (%)	Sludge (%)	Salt cake (%)	Composite (%)	Sludge (%)
Anions	Nitrate	25.37	27.80	66.43	60.56	49.26
	Nitrite	17.62	3.82	1.27	0	0
	Phosphate	3.23	13.79	1.04	1.11	0.97
	Hydroxide	4.41	3.39	0.71	0.27	0.27
	Carbonate	0.57	1.88	0.32	4.28	10.71
	Sulfate	0.00	1.02	0.98	1.14	1.04
	Cyanide	0.00	0.43	0	0	0
	Chloride	0	0.15	0	0.67	0.56
	Fluoride	0	0.12		0.12	0.12
Metals	Na, K	32.70	29.25	28.13	23.8	19.95
	Al	6.88	2.11	0.51	0.31	0.9
	Cr(III)	0	3.82	0		
	Fe	0	1.38	0	0.13	0.36
	Bi	0	0.52		0	0
	Ca, Mg	0	0	0.88	5.8	11.7
	Cr(VI)	0.70	0	0		
	Zr	0	0.43	0	trace	
	Cd	0.13	0.12	0.88	trace	
	Cr			0	0.01	0.04
	Sr	0	0.11	0	0.02	0.04
	W	0	0.11	0	0	0
	Carbon	0.28	0	0		
	U	<1	<1	<1	1.02	1.46
	Th				0.51	1.5
Radionuclide activity, Ci/L		0.83	2.5	0.13	0.055	0.15

The radioactive components of the Hanford and ORNL tank wastes are predominantly cesium and strontium, with lesser amounts of lanthanide elements and much lower activities of transuranium elements. The combined volume of the DST and SST tanks is 2.3×10^8 liters of liquid and solid waste with 240 million curies of radioactivity. The data for the Hanford SSTs indicate the contents (without water) to be 200×10^6 kg of chemicals with only 1.5×10^6 kg of radioactive elements. Separation of this 1% radioactive content from the total solid waste mixture would certainly minimize the quantity of DOE tank wastes that must be disposed of as high-level radioactive wastes - a prime example of how separation technology can be used for waste minimization.

FY 1993 WASTE MANAGEMENT PLANS FOR DOE SITES

Current DOE waste management strategies vary from site to site, depending on the local attitudes and available expertise at the particular site. Much of the planning at each site appears to be based on the idea that the waste should eventually be moved to some undefined government waste disposal site.

As previously discussed, the Idaho site has converted large amounts of acid raffinate waste to oxide forms, successfully preparing a stable waste form from reprocessing of spent naval nuclear fuels. Suspending the calcining process at Idaho prevented the site from completely eliminating the potential hazards associated with storing the liquid acidic waste. However, there is still time to complete the calcination process, which would allow safer storage of the waste until a final DOE repository is available.

At the Savannah River site, DOE has built a large vitrification plant to convert the tank waste into glass logs for shipment to the proposed long-term storage facility. The intended start-up time has been extended to 1994,[8] and several pretreatment separations processes are being used to prepare an acceptable feed to the vitrification plant.

Since 1985, Oak Ridge has planned and proposed to build a waste handling packaging plant (WHPP) to compact the site's TRU solid waste and, in a secondary process, solidify the MVST waste into remote-handled TRU salt cake.[9] The salt cake would then be shipped to the Waste Isolation Pilot Plant (WIPP) in New Mexico. Problems with this plan include the uncertainty of the WIPP receiving the remote-handled TRU waste and the need to incorporate waste minimization planning.

The ORNL Waste Management plan to build this facility and convert the MVST waste to salt cake calls for this process to be operational in 2007. In the meantime, the capacities of the MVST tanks have been approached several times. Additional space for the tank wastes has been provided three times by removing 189 m^3 (50,000 gal) of portions of supernatant and liquid from the tanks and converting it to a grout. Chemical analyses of the MVST wastes show that the TRU elements are in the sludge, in equilibrium with an essentially non-TRU supernatant liquid. However, the liquid is contaminated with cesium-137 and strontium-90. The conversion of the supernatant liquid to a grout has included no separation of radioactive components from the supernate and has increased the waste volume by about 50%. The grout was formed in concrete casks, and the grout-filled casks were stored on the ORNL site. A major

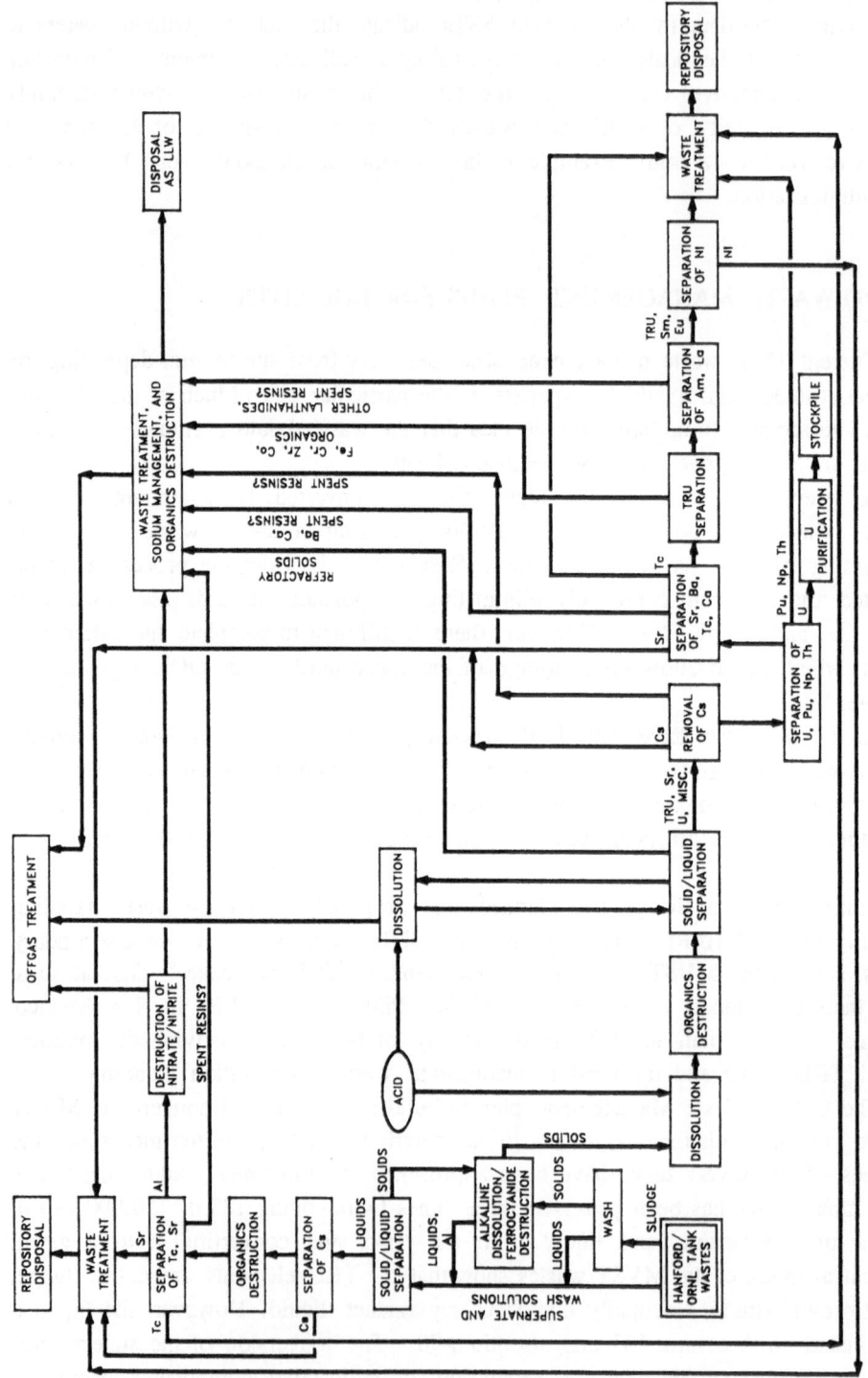

Figure 1. Functional flowsheet for Hanford/ORNL tank waste pretreatment:
Option 3 (late uranium separation)

concern with the ORNL Waste Management plan is the uncertainty of continued approval and acceptance of the grouting of supernatant liquid by the state of Tennessee.

Another strategy to prevent overfilling of the ORNL MVST containment has been evaporation of the supernatant liquid. In-tank vaporization has been tried with limited success, and a proposed external vaporization unit near the tanks could provide 189 to 378 m^3 (50,000 to 100,000 gal) of new space. The major effect of the vaporization processes is removal of water, which also increases the percentage of solids in the tanks. All of the radioactivity remains in the tanks.

The ORNL Waste Management efforts to provide MVST space may be inadequate for continued site operation into the next century. Therefore, the Waste Management plan includes the phased construction of additional storage tanks in 1999, which would provide storage for an additional 1890 m^3 (500,000 gal) of TRU waste at an estimated cost of $48 million. The overall plan calls for increasing the MVST storage volumes from 1514 to 3404 m^3 (400,000 to 900,000 gal) by the time the WHPP is operational (2007).

The Waste Management planning for tank waste at ORNL is also beginning to consider other options which include chemical separations and waste minimization. Without minimizing the MVST wastes through separation of the hazardous components, the projected volume of salt cake from the total conversion of the present MVST contents is 378 m^3 (100,000 gal) of salt cake for shipment to New Mexico's WIPP. A chemical separations pretreatment could conceivably minimize the volume of this hazardous TRU waste to less than 2 m^3 (500 gal). The cesium-137 and strontium-90 contents of the MVST wastes can be separated by ion-exchange technologies, and these shorter-lived isotopes can be safely stored as a much smaller volume of solid radioactive waste. The other radioactive and hazardous elements can also be separated by a variety of established technologies to achieve a relatively small volume of hazardous waste for storage.

As already mentioned, the Hanford site's waste tanks were defined in 1991 by Assistant Secretary of Energy Duffy as the largest and most hazardous problem for DOE. Waste management planning for the Hanford tank waste has been under way since the early 1980s,[10,11] but the plans for handling this waste have vacillated from "in situ" fixation to removal of the (solidified) tank waste completely out of the state of Washington.

A dramatic change in the Hanford waste management plan came with the Tri-Party Agreement.[12] Specifications of that Agreement in 1989 included (1) stabilization of the SSTs; (2) characterization by analysis of all contents of the tanks; and (3) solidification (by grouting or vitrification) of the total tank contents, followed by shipment of the vitrified waste to the DOE long-term storage facility. The first requirement was needed to prevent additional leakage of hazardous wastes from the SSTs. The other two specifications of the Agreement are difficult to defend and may be technically unachievable. In particular, removal and vitrification of all the 230 million liters in the tanks can only be accomplished at unbearable expense to the citizens of the United States. The Hanford cost estimate to convert all of the tank wastes into 200,000 glass logs[13] and to store the logs in the proposed DOE long-range repository is at least $750 thousand per log, or a total of $150 billion. (Recent estimates by the Savannah River and Hanford sites for waste vitrification have increased to $1 million per log.[14])

The Tri-Party Agreement specification requiring tank characterization to determine the exact chemical compositions of the tanks is technically unrealistic because of the heterogeneity of the contents of the tanks. Such total characterization of all the tanks is unnecessary for the proposed vitrification of the wastes and also for pretreatment of the waste before vitrification. A recent General Accounting Office (GAO) report[15] calls attention to the unreasonable extreme of this complete characterization. A more useful and much more economical characterization plan would be to accept the present analytical data as adequate until the wastes are retrieved or removed and placed into a fuel tank for processing, where the wastes can be homogenized and then analyzed. After homogenization of the wastes, characterization can be done to better assess the feasibility of various pretreatments and/or vitrification.

The recent GAO report[15] recommends that DOE delay the vitrification plant at Hanford and review the Tri-Party Agreement on Hanford waste management to arrive at a fair and realistic plan. Hanford and DOE are considering some alternate strategies that include pretreatment of the waste to separate the radioactive components into low-level and TRU wastes. The amount of pretreatment and the relative amounts of the two resulting wastes will be very important parameters to determine, and the Hanford Tank Waste Remediation Systems (TWRS) effort is considering many options. Barker et al.[16] have estimated that appropriate pretreatment of the Hanford tank waste can reduce the final volume of high-level waste to approximately 25,000-40,000 glass logs. The greatest high-level waste minimization could result from the CLEAN philosophy option,[17] which proposes to undertake the study of technologies with the goal of minimizing the Hanford tank waste material that must be placed in long-term storage to only 1000 glass logs.

The results of various pretreatment scenarios to reduce the number of vitrified logs of Hanford tank wastes and the related costs are shown in Table 3. "No pretreatment" means that the total volume of all 177 tanks is vitrified. The "sludge wash" minimum pretreatment scenario corresponds to dissolution of largely nonradioactive soluble components from the sludge waste into water at a pH of 10 to 14. This supernatant liquid would be decontaminated and then converted into a low-level waste form. The remaining sludge would be vitrified into about 40,000 glass logs. At a cost of $750 thousand per glass log, this would amount to ~$30 billion. An additional cost to this pretreatment would be the sludge washing steps and management of the supernatant liquid. Such additional costs have not been determined, but they are generally believed to be considerably less than the vitrification costs.

Further studies from the Hanford TWRS have indicated that the application of pretreatment separations to the washed sludge can minimize the Hanford tank waste volumes that must be vitrified and placed in the long-range repository to 12,000 logs.[13] A recent Pacific Northwest Laboratory suggestion of an ultimate separations pretreatment scheme (the CLEAN philosophy) was discussed at the 1992 Actinide Separations Conference by a group of chemical separations specialists. Their conclusion was that it is technically feasible that extensive application of chemical separations pretreatment of the Hanford tank wastes can reduce the volume of high-level waste that requires long-term storage to only 1000 vitrified logs.

Table 3. Effects of chemical separation pretreatment on the volumes of high-level TRU waste from the Hanford and Oak Ridge underground storage tanks.

Effects of Pretreatment on Volume and Cost of Hanford Vitrified Glass Logs

Pretreatment (Ref. No.)	Number of Logs	Costs ($750,000/Log)
None (13)	200,000	150 Billion
Sludge wash (13)	40,000	30 Billion
TWRS (13)	12,000	9 Billion
Ultimate − Clean (17)	<1,000	1 Billion

Effects of Pretreatment on Volume of ORNL MVST TRU Wastes

Pretreatment (Ref. No.)	Waste Form (Salt Cake)
None (4)	400 m^3 (100,000 gal)
Sludge wash − remove all solution[a]	80 m^3 (20,000 gal)
Only radioactive components[a]	2 m^3 (<500 gal)

[a]Calculations herein.

WASTE MINIMIZATION FOR THE DOE TANK WASTES

The basis for better waste management of the tank waste must be the earnest adoption, by DOE and its contractors, of a waste minimization philosophy and approach for treatment of all DOE radioactive tank wastes. The simple 1-2-3 philosophy of (1) separation of the hazardous from the nonhazardous waste materials, (2) recycling of useful chemicals and materials, and (3) disposing of only the minimum remaining volumes of radioactive and other hazardous wastes will provide more rapid, more feasible, and much more economic solutions to the tank waste problems. This waste minimization philosophy excludes all "fix-it-quick" schemes for management of DOE underground storage tank wastes. Current plans based on those attitudes need to be replaced with plans to (1) eliminate immediate hazards and (2) minimize the volumes of wastes that must be stored for 10^4 years. To do this, the best means must be identified to separate and treat the waste so it will be in forms that offer little or no hazard at any time in the future. Furthermore, this must be done at costs that are acceptable to the U.S. taxpayer. This may mean that the waste will not be placed in its final waste forms "immediately;" instead, better separation and treatment methods may need to be developed and tested. In any case, DOE's long-range waste repositories should certainly be reserved for the wastes having the greatest potential for long-range environmental insults. A beneficial waste acceptance criterion for the long-range repository would be certification that pretreatment separations have been applied to the extent possible to minimize the high-level waste disposal, thus conserving the valuable repository space for future generations.

A variety of well-developed separation techniques are available and are technically feasible for application to tank wastes. These include solid-liquid separations, solvent extraction, ion exchange, membranes, calcination, and absorption techniques. A combination of appropriate technologies will likely prove most useful for achieving the most effective waste minimization.

The application of classical chemical processing technology to a particular problem generally evolves through several steps: (1) developing the basic chemical reaction scheme, (2) proving the chemical technology scheme through lab testing, (3) expanding the testing to engineering scale, (4) testing the technology in a shielded environment by processing real waste, and (5) proving the integrated technology in a pilot-scale operation utilizing real waste to provide design data for full-scale plant operation. Many of the chemical separations technologies applicable to the DOE radioactive tank wastes are well developed through the third step. Therefore, an effective approach to establishing appropriate pretreatment schemes for the tank waste management would be to proceed with steps 4 and 5 for some of these partially developed technologies. Based on those results, DOE could then establish appropriate large-scale operations for separating the tank wastes. Delay of any large-scale operations until the pilot-plant testing is accomplished would ensure that the most effective and practicable waste minimization scheme is identified and that design data are available. As of 1993, no pilot-scale testing of separations technologies is being done for waste minimization of DOE tank wastes.

The DOE complex is considering various degrees of pretreatment of tank waste to minimize the resulting volume of waste product that must be stored in a long-range repository. This minimization can save the U.S. government billions of dollars. Figure 1 presents an integrated flowchart with the CLEAN philosophy for separation and pretreatment of DOE Hanford and Oak Ridge tank wastes to minimize the amount of TRU waste that must be stored for $>10^4$ years. This flowchart is generic with respect to separations technologies; but each flowsheet operation is technically feasible, and a number of operations have at least one available technology. An acid-recycle process is not shown but is also an available technology.

The development of pilot-plant operations for the technologies identified in this flowchart can lead DOE to the most appropriate waste minimizing technologies for application to the management of wastes in the underground storage tanks. Estimates of volumes of vitrification product resulting from various degrees of pretreatment of the Hanford tank waste are summarized in Table 3. The optimal degree of pretreatment can be determined from the pilot-plant operations. The important point is that waste can be minimized by the 1-2-3 philosophy discussed here. Also provided in Table 3 are estimates of the effects of pretreatment on the volume of remote-handled TRU waste in the ORNL MVST tanks. The current ORNL Waste Management plan to convert these wastes to a salt cake form would produce 400 m^3 (10^5 gal) of remote-handled salt cake for shipment to the WIPP site. Since the "sludge-wash" pretreatment reduces the volume of vitrified logs by a factor of 5, and since the volume of logs for the non-pretreatment of the wastes is nearly the same volume of the same waste in salt-cake form, we assume that the sludge wash pretreatment of the MVST waste can reduce this ORNL TRU waste from 400 m^3 to 80 m^3. Extensive pretreatment of the ORNL MVST waste could reduce the ORNL waste volume for ultimate disposal to only 2 m^3 (500 gal). Again, the optimal pretreatment scheme can be determined from pilot-plant operations using the 1-2-3 waste minimization philosophy.

REFERENCES

1. "Progress in Nuclear Energy. Series III: Process Chemistry, Volume 2," F. R. Bruce, J. M. Fletcher, and H. H. Hyman, eds., Pergamon Press (1958).

2. H. O. Weeren, "Slucing Operations at Gunite Waste Storage Tanks," ORNL/NFW-84/42 (1984).

3. T. Tamura and H. O. Weeren, "Disposal of Waste by Hydraulic Fracturing," presented at the 2nd Hazardous Material Conf., Philadelphia, June 6-8, 1984, published in Proc. 2nd Hazardous Materials Conf., pp. 483-97 (1984).

4. T. E. Myrick et al., "The Emergency Avoidance Solidification Campaign of Liquid Low-Level Waste at Oak Ridge National Laboratory," ORNL/TM-11536 (March 1991).

5. Private discussion with A. Olsen: The Idaho site has recently restarted the calcination system to treat the remaining acid waste (May 1993).

6. (a) "Defense Waste Management Plan," DOE/DP-0015, U.S. Dept. of Energy Headquarters (1983); (b) M. D. Boersma, "Process Technology for the Vitrification of Defense High-Level Waste at the Savannah River Plant," ANS, Fuel Reprocessing and Waste Management Proceedings, 1, pp. 131-47 (1984).

7. Approximately 500 Ci/liter is based on a spent fuel that has cooled for 1 year with a total activity of 1.72×10^6 curies per ton of initial uranium and a process feed concentration of 250 g/liter.

8. "High-Level Waste System Plans," Rev. 0, WER-DPM-92-0153 (December 1992).

9. D. W. Turner, J. B. Berry, and J. W. Moore, "An Overview of the ORNL Waste Handling and Packaging Plant," Waste Management '90, published in the proceedings (February 1990).

10. K. S. Murthy, L. A. Stout, B. A. Napier, A. E. Reisenauer, D. K. Landstrom, "Assessment of Single-Shell Tank Residual Liquid Issues at Hanford Site, Washington," PNL-4688, Pacific Northwest Laboratory, Richland, Washington (1983).

11. DOE/EIS-0113, "Final Environmental Impact Statement: Disposal of Hanford Defense High-Level, Transuranic, and Tank Wastes, Hanford Site, Richland, Washington," U.S. Department of Energy—Headquarters, Washington, DC (1987).

12. Federal Facility Agreement and Consent Order (May 1989).

13. J. T. Bell Participation: (a) Independent Engineering Review of Hanford Tank Waste Management Program, November 18-22, 1991; (b) Hanford Tank Waste Remediation Technology Workshop, February 11-12, 1992; (c) TWRS Technology Working Groups National Workshop, June 29-June 1, 1992.

14. Private communication with Frank Graham, Actinide Separations Workshop (May 1993).

15. General Accounting Office, "Nuclear Waste, Hanford Tank Waste Program Needs Cost, Schedule, and Management Changes," GAO/RCED-93-99 (1993).

16. S. A. Barker, C. K. Thornhill, and L. K. Hilton, "Pretreatment Technology Plan," WHC-EP-0629, Westinghouse Hanford Company, Richland, Washington.

17. J. L. Straaslund et al., "Clean Option: An Alternative Strategy for Hanford Tank Waste Remediation," PNL-8388, Vol. 1, UC T21 (1992).

CHEMICAL PRETREATMENT OF SAVANNAH RIVER SITE

NUCLEAR WASTE FOR DISPOSAL

D.T. Hobbs and D.D. Walker

Westinghouse Savannah River Company
Savannah River Technology Center
Aiken, South Carolina 29808

INTRODUCTION

The Savannah River Site (SRS), located near Aiken, SC, is presently the nation's primary source of nuclear materials for defense, space, medical, and energy applications. The site, which was built in the early 1950's, comprises over 750 square kilometers of land with extensive support facilities. Nuclear materials produced at SRS are generated in heavy-water reactors by irradiating appropriate target materials with neutrons from uranium fuels and then chemically separating the products in two on-site processing plants.

Since start-up over 400 million liters of high level waste have been generated. This waste has been evaporated to about 130 million liters, which is now stored in large underground tanks. The principal radionuclides in the waste are ^{90}Sr and ^{137}Cs. Approximately 10% of the waste consists of iron, manganese, uranium and aluminum oxides and hydroxides, which are precipitated upon neutralization of the acidic nitric acid wastes produced in the chemical separation operations. These solids are referred to as sludge. The sludge contains most of the strontium and small amounts of actinides not recovered in the reprocessing plants. The remainder of the waste is liquid and salt cake, which consists primarily of sodium nitrate, sodium nitrite, sodium aluminate, and sodium hydroxide. This waste contains ^{137}Cs and traces of other soluble radionuclides.

The HLW will be pretreated to reduce the amount of waste which is disposed of in the high-cost borosilicate glass wasteform. This pretreatment will concentrate the radionuclides in a small volume for vitrification. The remaining high-volume, low-level waste will be disposed of in a cement wasteform referred to as Saltstone. A schematic diagram of the pretreatment and final disposal of HLW at the SRS is provided in Figure 1.

HLW pretreatment operations, referred to as In-Tank Processing, will be carried out in waste tanks which have been modified for chemical processing. In-Tank Processing consists of two separate processes, Extended Sludge Processing and In-Tank Precipitation. A schematic diagram of the Extended Sludge Processing operation is provided in Figure 2. The Extended Sludge Processing operation will prepare the sludge fraction of HLW for feed to the Defense Waste Processing Facility (DWPF) for vitrification. A schematic diagram of the In-Tank Precipitation Process is provided in Figure 3. The In-Tank Precipitation Process will remove greater than 99.9% of radioactivity in the salt fraction of HLW into a solid phase which will be combined with sludge and vitrified. The decontaminated salt solution will be disposed of in a cement wasteform, Saltstone.

Chemical Pretreatment of Nuclear Waste for Disposal, Edited by
W.W. Schulz and E.P. Horwitz, Plenum Press, New York, 1995

17

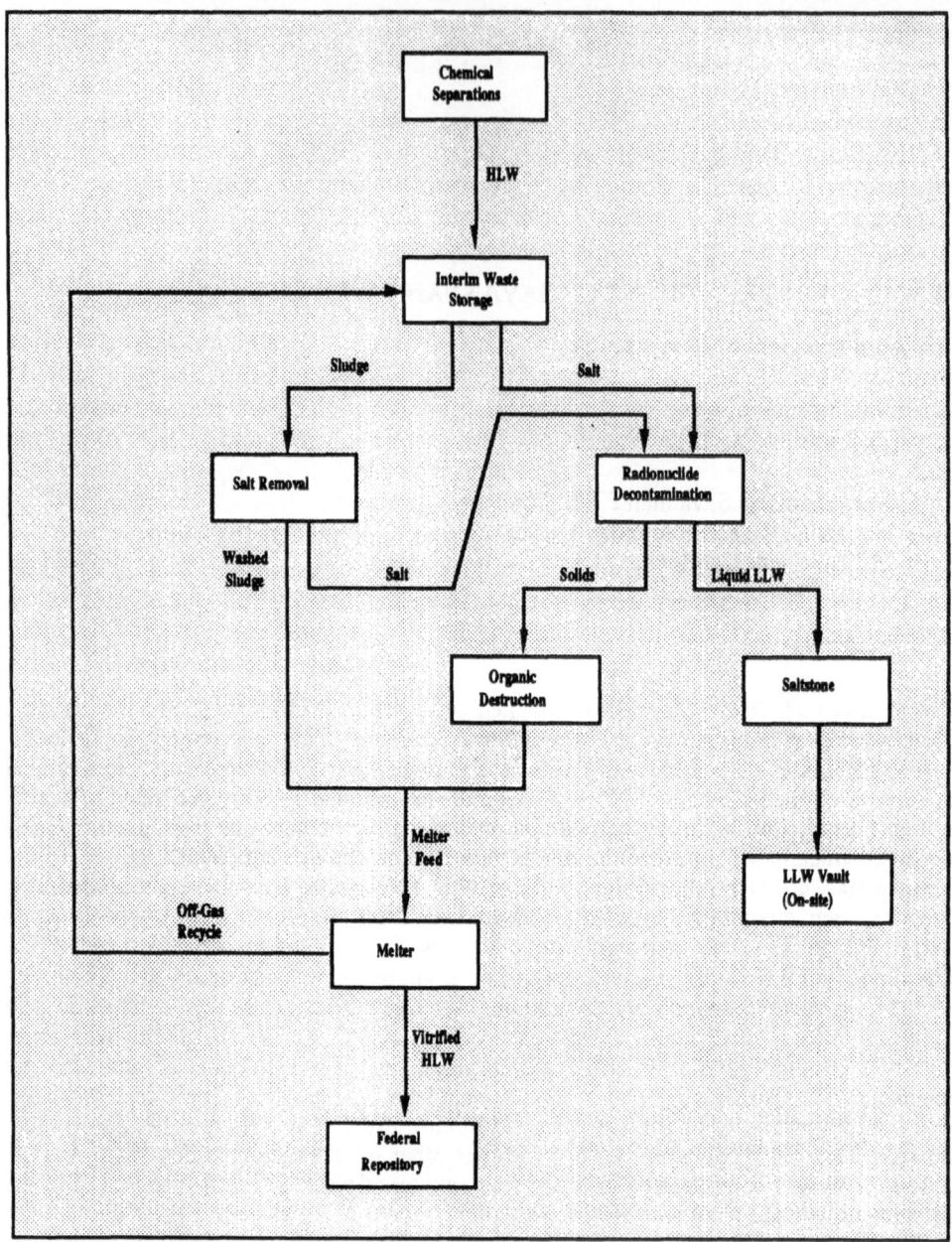

Figure 1. Schematic Diagram for Pretreatment and Disposal of High-Level Waste at the Savannah River Site.

EXTENDED SLUDGE PROCESSING

The purpose of the Extended Sludge Processing (ESP) operation is to reduce the solids content of the sludge and ensure its compatibility with the borosilicate glass wasteform. Reducing the soluble salts content results in significant cost savings by reducing the total amount of glass produced in the DWPF. The soluble salts will be reduced by a series of batch washings. A schematic flow diagram of the ESP operation is presented in Figure 2.

Sludge processing will occur in three 4.9 million-liter, HLW storage tanks which have been modified for processing. Sludge wastes will be removed from storage tanks and transferred into one of two tanks in which the sludge is washed concurrently. The sludge is contacted with wash water for a period of time to ensure a homogeneous mixture, the agitation is then stopped, and the

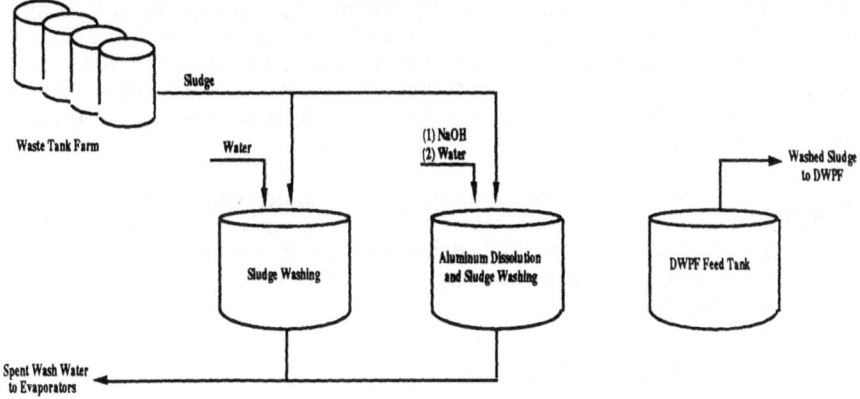

Figure 2. Schematic Diagram of the ESP Operation.

solids allowed to gravity settle. The liquid is then decanted from the settled solids. This series of steps is repeated until the soluble salt content in the liquid phase is reduced to ≤ 2.5 weight percent of the total solids content. Wash solutions produced in ESP will be transferred to an evaporator system or reused in dissolving saltcake, depending on the salt concentration. The wash water from the first tank will be reused to wash sludge in the second tank.

For sludge wastes which have a high aluminum content, an additional step is added to dissolve the aluminum prior to water washing. The sludge is heated with sodium hydroxide to dissolve approximately 75% of the aluminum. After dissolution, the tank agitation is stopped, the undissolved sludge solids are allowed to gravity settle, and the supernatant liquid is decanted. The remaining sludge will then be washed as described above. Aluminum dissolution will be carried out only in one of the three processing tanks.

After washing has been completed, both batches of sludge will be combined into one tank for storage. From this tank, washed sludge will be transferred to the DWPF through underground transfer lines. The next batch of sludge for the DWPF will then be produced using the two empty tanks.

ALUMINUM DISSOLUTION CHEMISTRY

One of the major constituents of SRS HLW is aluminum. Aluminum is used for cladding and alloying materials in fuel and target assemblies. If left in the sludge, aluminum adversely affects glass viscosity and adds to the volume of waste processed in the DWPF. Aluminum is present in both the supernatant liquid and in the sludge wastes. In the supernatant liquid, aluminum is present as the aluminate ion, $Al(OH)_4^-$. In sludge, aluminum is present in three forms, alumina trihydrate or gibbsite, alumina monohydrate or boehmite, and sodium aluminosilicate. Only the gibbsite is readily soluble at the conditions which will be used in the ESP operation. The chemical reaction for the dissolution of gibbsite is given below:

$$Al_2O_3 \cdot 3H_2O_{(s)} + 2NaOH_{(aq)} \quad ---> \quad 2NaAl(OH)_{4(aq)}.$$

To dissolve the gibbsite, 50 wt.% sodium hydroxide solutions will be added to the sludge slurry in the processing tank to provide a minimum initial ratio of three moles of free hydroxide per mole of acid-soluble gibbsite, and a final liquid phase free hydroxide concentration of 3 molar. The sludge slurry will be steam heated to between 80 and 90°C and agitated until all of the gibbsite dissolves. It is estimated that it take approximately three days for the gibbsite to dissolve.

FULL-SCALE ESP DEMONSTRATION

A full-scale demonstration of ESP was carried out in 1982-1983. Approximately 473,000 liters of sludge were successfully transferred a distance of about 3.2 kilometers from the storage tank in F-Area to the processing tank in H-Area via the inter-area transfer line. The sludge was slurried in the processing tanks using three slurry pumps. This sludge had a high aluminum content, and therefore, was treated with sodium hydroxide to dissolve the gibbsite. Seventy-nine percent of the total aluminum was removed by aluminum dissolution. After aluminum dissolution, the sludge was washed with three consecutive batches of wash water which reduced the soluble salt content to the DWPF feed requirement of ≤2.5 wt.%. The final solids content of the settled sludge was 14 wt.%.

The slurry pumps used in the demonstration are variable-speed centrifugal pumps with a flow capacity of 250 liters/sec and a nominal mixing radius of 12f meters powered by a 224 kJ/sec motor. The pump is constructed of type 304L stainless steel with graphite sleeve bushings for the bottom seal. Since the demonstration, the design of the mechanical seals has been changed to decrease the bearing water leak rate.

IN-TANK PRECIPITATION PROCESS

The purpose of the In-Tank Precipitation (ITP) process is to concentrate the radio-cesium and radio-strontium into a small-volume solid phase which can be separated from the liquid phase and vitrified with the washed sludge in the DWPF. In this process (see Figure 3), sodium tetraphenylborate and monosodium titanate are added to a waste tank containing redissolved salt and supernatant liquid. The cesium and potassium precipitates as cesium tetraphenylborate and potassium tetraphenylborate. The strontium is adsorbed by the monosodium titanate, which remains a solid throughout the process. The supernate, with most of its radioactive constituents now in solid form, is then filtered through sintered stainless steel filters with a 0.5 μm pore size.

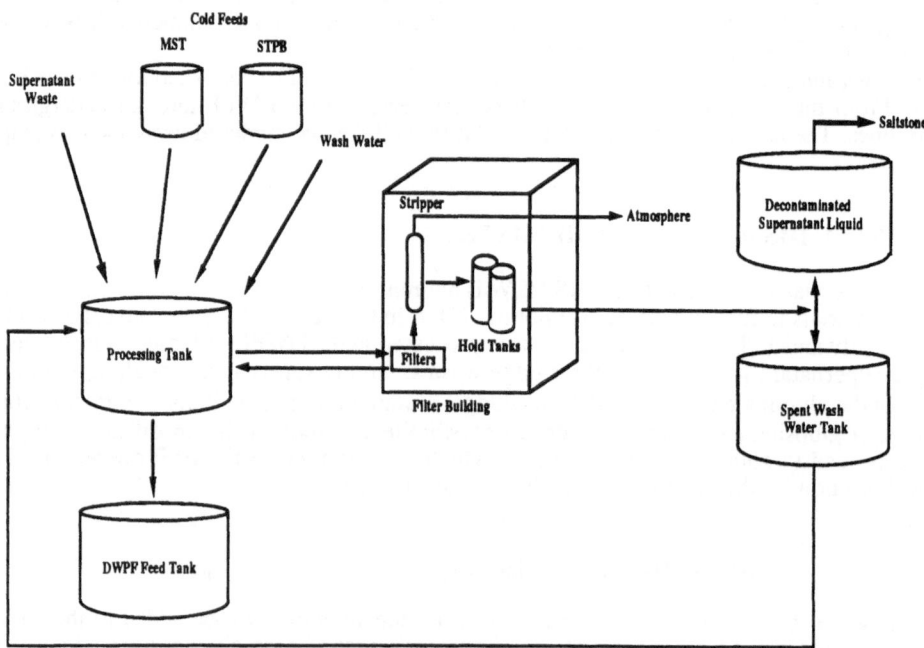

Figure 3. Schematic Diagram of the In-Tank Precipitation Process

The decontaminated supernate filtrate will be disposed of as a low-level radioactive waste in a cement wasteform, Saltstone. The cement wasteform will be placed in an engineered vault onsite. The precipitate, which is unable to pass through the filters, is recycled back to the waste tank. This material is then concentrated and washed to remove residual salts. The washed precipitate, containing virtually all the radioactivity of the original supernate, will be incorporated into borosilicate glass in the DWPF.

ITP PROCESS CHEMISTRY

In the precipitation process, tetraphenylborate anion is added as the water soluble sodium tetraphenylborate. The waste supernate contains about 2.5×10^{-4}M cesium and 0.02M potassium, and both are precipitated as the respective tetraphenylborate (TPB) salts. Over 98% of the Cs^+/K^+ precipitate is the potassium salt. The precipitation reaction can be represented as:

$$B\emptyset_4^- \text{ (aq)} + Cs^+/K^+ \text{ (aq)} \longrightarrow Cs/K \, B\emptyset_4 \text{ (s)}$$

The effects of several experimental variables on cesium removal via sodium tetraphenylborate precipitation have been studied. These variables include supernate feed composition [$(Na^+) = $ 2.8-6.2M, $(OH^-) = $ 1-3.8M, cesium activity = 0.02-1.0 Ci/L], sodium tetraphenylborate excess, contact time, and purity of sodium tetraphenylborate. Sodium tetraphenylborate was added as a 0.5M solution in 0.01M sodium hydroxide. The decontamination factor (DF) for cesium ranges from 10^4 to 10^6 depending on the cesium and sodium concentrations in the feed. The expected DF values for cesium in salt supernate can be calculated from the solubilities of cesium and sodium tetraphenylborate. The calculated DF values agree well with values measured on actual supernate.

The cesium DF is affected strongly by the sodium and potassium ion concentrations. The concentration of hydroxide ion and the radioactivity level have only a minor effect on cesium DF. With a sodium tetraphenylborate excess of 0.015M, a high cesium DF ($>10^5$) is always obtained when the supernate contains less than 3.5M sodium ion. However, the solubility product for sodium tetraphenylborate is exceeded when supernate containing more than 4.5M sodium is used. The cesium DF then varies with contact time because most of the tetraphenylborate anion added is precipitated at the high sodium concentration and is not available immediately to precipitate cesium. Sodium tetraphenylborate solubility decreases from about 0.9M in water to 0.00011M in supernate containing 7M sodium. A high cesium DF can be obtained if enough excess sodium tetraphenylborate is added and a longer contact time is used to allow the cesium to exchange with the sodium.

ADSORPTION OF STRONTIUM AND ACTINIDES BY MONOSODIUM TITANATE

Most of the strontium in the HLW is precipitated as strontium hydroxide in the sludge. The remaining concentration of soluble strontium in the supernate is about 2×10^{-7}M. Strontium can be removed from the high sodium ion solution by the addition of monosodium titanate or MST ($NaTi_2O_5H$). The MST is an inorganic ion-exchanger developed by R. Dosch and coworkers at Sandia National Laboratory in the mid-1970's.

In alkaline solution, removal of strontium is reported to be a surface ion-exchange process between the sodium in the MST and the strontium in solution. The chemical reaction for this reaction is shown below;

$$NaTi_2O_5H_{(s)} + Sr^{2+}_{(aq)} \longrightarrow Sr(Ti_2O_5H)_{2 \, (s)} + 2Na^+_{(aq)}.$$

In laboratory tests with excellent mixing, equilibrium is achieved in 48 hours upon contact of a test solution with the MST. Greater than 99% of the strontium is removed when a solution containing 2×10^{-7}M strontium is treated with 0.5 grams of MST per liter of solution.

Like strontium, most of the plutonium and uranium are in the sludge. However, small amounts of plutonium and uranium are soluble in the concentrated alkaline salt solutions. Tests have shown that both plutonium and uranium are removed from solution upon addition of MST. Greater than 90% of the plutonium and approximately 30% of the uranium are removed. However, because of the higher solubility of uranium, the loading of uranium onto the MST is larger than that of plutonium.

The mechanism of the removal of plutonium and uranium by the MST is not known conclusively. Preliminary results indicate that the removal of uranium may be a surface adsorption mechanism, whereas plutonium may be removed by either adsorption or ion-exchange or both. The removal of strontium and plutonium increases with temperature. In contrast, the removal of uranium decreases with increasing temperature. From competition experiments, the removal of plutonium and strontium are not affected by the presence of each other, but are decreased by the presence of uranium.

Because of the potential for the accumulation of fissile material by adsorption onto the MST, an assessment of the nuclear criticality safety during the ITP process has been carried out. The results of the analysis indicate that there is no credible potential for criticality. Because of the low solubility and low mass loading onto the MST, the amount of MST required to load a sufficient quantity of fissile material for a criticality incident is more than that required to process all of the liquid HLW in the ITP process.

CHEMICAL AND RADIOLYTICAL STABILITY OF POTASSIUM AND CESIUM TETRAPHENYLBORATE PRECIPITATE

The precipitation process can be performed in existing waste tanks only if the K/Cs TPB precipitate remains chemically and radiolytically stable long enough to filter and wash it. Tests indicate that the precipitate is chemically stable in the alkaline salt solutions which will be produced during precipitation and washing. In the absence of a radiation field, the tetraphenylborate salts are stable for months in alkaline solutions and slurries at temperatures up to 70°C.

The tetraphenylborate salts will hydrolyze under acidic conditions. However, acidic conditions are not produced during any stage of the HLW pretreatment operations carried out in the waste tanks. Acid hydrolysis will be utilized in the DWPF to decompose the precipitate and allow separation of the organic carbon from the remaining aqueous phase by steam distilling.

Gamma irradiation tests indicate that the insoluble K/Cs TPB precipitate is about 15 times more stable than sodium tetraphenylborate. At an expected dose rate of 2.7 J/kg-s, approximately 10% of the K/Cs TPB will decompose during one year of storage. Radiolytic decomposition products include in decreasing yield; benzene (50%), phenol (25%), biphenyl (20%), and phenylboric acid (5%). The effects of K/Cs tetraphenylborate decomposition on the salt DF can easily be overcome by the presence of a small excess of sodium tetraphenylborate during the expected processing period. Degradation of the precipitate is still to be avoided for flammability, processing, and environmental considerations.

FILTRATION OF ITP SLURRIES

Cross-flow filtration was chosen for separation of the solid tetraphenylborate and sodium titanate from the decontaminated supernate (see Figure 4). Cross-flow filtration is recommended with low solids loading, relatively small particle size, and in cases where power costs for pressure operation are not a problem. During filtration the pressurized precipitate slurry is pumped down the filter tubes at high velocities. The filtrate weeps through the pores and is collected in the annular space between the filter tube and its housing.

The high linear velocity of the slurry helps sweep the filter surface clean and maintain high filtrate flux. The slurry is recycled to the precipitation/feed tank until the desired solid concentration is reached. The filter tubes can be backpulsed with air or other inert gas to remove solids any solids which have collected in the filter pores. In addition, a filter cleaning procedure has been developed for when a sufficient amount of solids has accumulated in the filter pores to result in a significant decrease in filtrate flux or very high pressure drops across the filter. When this occurs, the filter is first flushed with water, and then with a 2-4 wt.% solution of oxalic acid. Oxalic acid dissolves solids in the filter pores, thereby reopening filter channels and restoring the desired filtrate flux and pressure drop across the filter. Flushing with sodium hydroxide solution has also been shown to be effective in cleaning plugged filters.

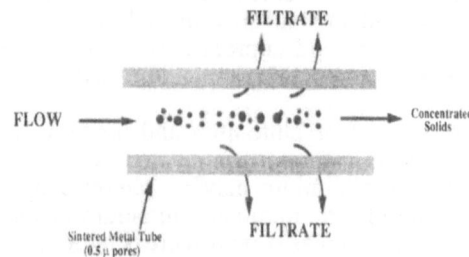

Fig. 4. Schematic Drawing of Cross-Flow Filtration.

FLAMMABILITY HAZARDS

There are two flammability hazards associated with the ITP process; (1) combustion of dry K/Cs TPB solids and (2) ignition of flammable vapor mixtures of benzene and hydrogen. Liquid slurries, which are produced in the ITP process, are not ignitable because of their high water content. However, deposits of dry solids from the slurries are combustible and produce dense, black smoke when burned. As a result of the radiolytic decomposition of the K/Cs TPB precipitate and water, benzene and hydrogen will be produced during the ITP process. Benzene is volatile and flammable, with a lower flammable limit (LFL) of 1.2% (v/v) in air. The LFL for hydrogen in air is 4.0% (v/v). Engineering and administrative controls have been developed to prevent the combustion of dry solids and the accumulation of flammable vapor mixtures. All of the waste tanks are equipped with active ventilation systems which continuously purge the vapor space of the tank. In addition, these tank, as well as the cold chemical NaTPB storage tank, are blanketed with nitrogen to reduce the oxygen content to below that required to support combustion. In the event of loss of power, backup diesel generators are available to provide electrical power to the ventilation system. In the event of loss of installed blowers, dedicated portable blowers can be installed.

FULL-SCALE ITP DEMONSTRATION

A full-scale demonstration of this process was conducted in 1983. In that demonstration 1.6 million liters of high-level radioactive supernate were transferred from Tank 24H to Tank 48H, treated with sodium tetraphenylborate and monosodium titanatefor three days, and then the precipitate was filtered. A total of 2.0 million liters of decontaminated supernate were produced. The average ^{137}Cs activity of the decontaminated supernate was 7400 d/m/ml. corresponding to a ^{137}Cs decontamination factor of 43,000. After filtration, about 150,000 liters of concentrated precipitate remained, which was water washed to remove sodium salts and reconcentrated to 10 wt.% solids.

CORROSION PREVENTION

The successful pretreatment of all of the HLW stored at the SRS requires the use of eight carbon steel waste tanks over the life of the ESP and ITP processes. To ensure that the service life will meet the planned schedule of about 15 years to pretreat all of the currently stored waste, the chemical composition of the wastes will be closely monitored and corrosion inhibitors added as necessary to prevent corrosion. Corrosion monitors are also being developed to provide on-line indication of corrosive conditions within the waste tanks.

Laboratory tests indicate that localized or pitting corrosion can occur during ESP and ITP operations which could render the tanks unusable. Pitting corrosion occurs as a result of the dilution or depletion of corrosion inhibitors originally present in the waste. Aggressive species in the waste which enhance the rate of attack include nitrate, sulfate, chloride and fluoride[1]. Nitrite and hydroxide have been shown to be effective inhibitors against pitting corrosion even in dilute salt solutions[2]. Minimum concentrations for each inhibitor have been established over the range of waste compositions expected for each process.

Stress cracking corrosion during aluminum dissolution was also identified as a possible corrosion concern because of the high hydroxide concentrations and temperature. Laboratory tests have measured the potential for stress corrosion cracking over a wide range of waste compositions that may occur during aluminum dissolution. These tests indicate that stress corrosion cracking will not occur if the hydroxide concentration is maintained below 8.0 molar, the temperature is less than 100°C, and there is a minimum nitrate concentration of 0.02 molar. These conditions provide a wide operating envelope for removing aluminum from wastes during the ESP operation.

SUMMARY

Two processes, Extended Sludge Processing and In-Tank Precipitation, have been developed and demonstrated at full-scale to pretreat the Savannah River Site High-Level Waste for permanent disposal. These processes will be carried out in waste storage tanks which have been modified for chemical processing. These processes will concentrate the radioactivity into a small volume for vitrification. The bulk of the waste will be sufficiently decontaminated that it can be disposed of as a low-level waste. The highly radioactive fraction will be vitrified in the Defense Waste Processing Facility. The decontaminated waste will be incorporated into a cement wasteform in the Saltstone Facility.

REFERENCES

1. D. F. Bickford, J. W. Congdon, and S. B. Oblath, "Corrosion of Radioactive Waste Tanks Containing Washed Sludge and Precipitate," <u>Materials Performance</u>, **27**(5), 16 (1988).
2. J. W. Congdon, "Inhibition of Nuclear Waste Solutions Containing Multiple Aggressive Anions," <u>Materials Performance</u>, **27**(5), 34 (1988).

ACKNOWLEDGEMENT

The information contained in this article was developed during the course of work under Contract No. DE-AC09-89SR18035 with the U.S. Department of Energy.

DISPOSAL OF HANFORD SITE TANK WASTES

M. J. Kupfer

Westinghouse Hanford Company
Post Office Box 1970
Richland, WA 99352

INTRODUCTION

Between 1943 and 1986, 149 single-shell tanks (SSTs) and 28 double-shell tanks (DSTs) were built and used to store radioactive wastes generated during reprocessing of irradiated uranium metal fuel elements at the U.S. Department of Energy (DOE) Hanford Site in Southeastern Washington state. The 149 SSTs (Figure 1), located in 12 separate areas (tank farms) in the 200 East and 200 West areas, currently contain about 1.4×10^5 m^3 of solid (predominantly) and liquid wastes. Wastes in the SSTs contain about 5.7×10^{18} Bq (170 MCi) of various radionuclides including ^{90}Sr, ^{99}Tc, ^{137}Cs, and transuranium (TRU) elements. The 28 DSTs (Figure 2) also located in the 200 East and West areas contain about 9×10^4 m^3 of liquid (mainly) and solid wastes; approximately 4×10^{18} Bq (90 MCi) of radionuclides are stored in the DSTs.

Important characteristics and features of the various types of SST and DST wastes are described in this paper. However, the principal focus of this paper is on the evolving strategy for final disposal of both the SST and DST wastes. Also provided is a chronology which lists key events and dates in the development of strategies for disposal of Hanford Site tank wastes. One of these strategies involves pretreatment of retrieved tank wastes to separate them into a small volume of high-level radioactive waste requiring, after vitrification, disposal in a deep geologic repository and a large volume of low-level radioactive waste which can be safely disposed of in near-surface facilities at the Hanford Site. The last section of this paper lists and describes some of the pretreatment procedures and processes being considered for removal of important radionuclides from retrieved tank wastes.

WASTE TANK SYSTEMS AND CONTENTS

Double-Shell Tank Wastes

The DSTs (tank-within-a-tank) were constructed from 1970 to 1985; all of the DSTs are designed to contain 3,800 m^3 of waste.

All the DSTs are constructed of mild steel (high carbon content). The sides and

Chemical Pretreatment of Nuclear Waste for Disposal, Edited by
W.W. Schulz and E.P. Horwitz, Plenum Press, New York, 1995

25

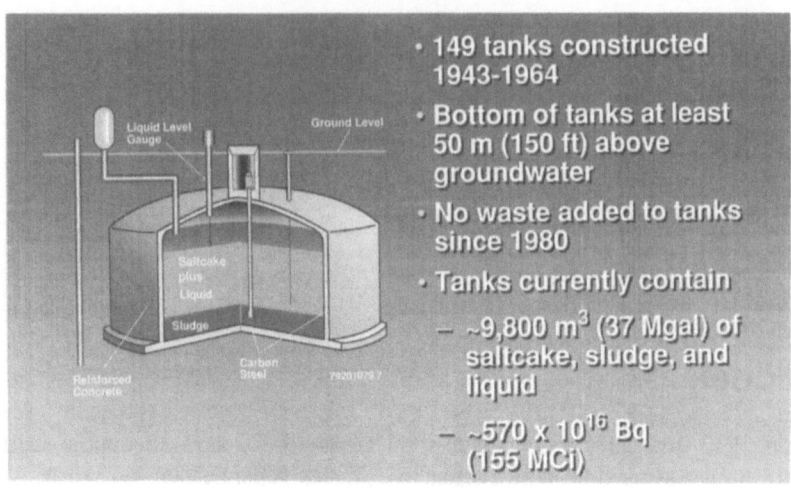

Figure 1. Single-Shell Tanks.

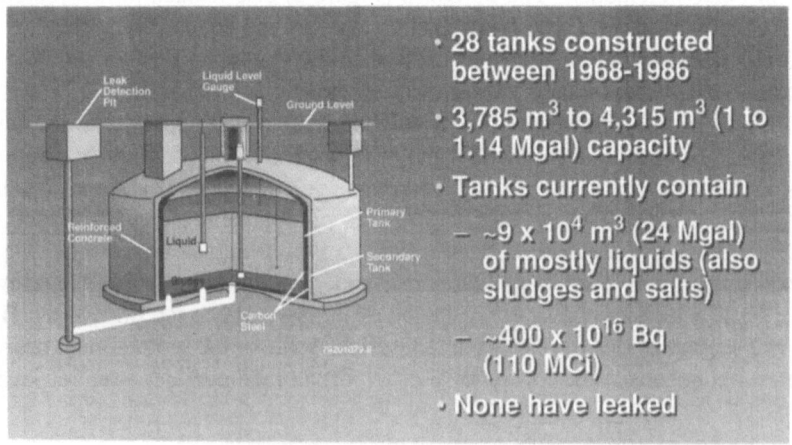

Figure 2. Double-Shell Tanks.

bottoms of all tanks are supported by concrete structures. Openings in the unsupported tank domes allow limited access for sampling wastes and for measuring waste temperatures and liquid levels. An extensive network of buried piping is provided for the transfer of liquid wastes and waste slurries within and between tank farms.

The total present and expected future inventory of wastes in the existing 28 DSTs is classified into five types:

- Neutralized current acid waste (NCAW)--5,300 m^3

- Neutralized cladding removal waste (NCRW)--3,300 m^3 of sludge

- Plutonium Finishing Plant (PFP) waste--970 m^3 of sludge

- Complexant concentrate (CC) waste--18,200 m^3

- Double-shell slurry (DSS), double-shell slurry feed (DSSF), and dilute noncomplexed (DN) wastes--75,700 m^3 (including the future addition of 35,300 m^3 DN waste to be evaporated to 5,300 m^3 DSSF)

All the DST wastes consist of a liquid portion and a solid portion. The NCAW is the waste that was produced when concentrated acidic plutonium-uranium extraction (PUREX) process high-level radioactive waste generated between 1983 and 1988 was made alkaline and stored in DSTs. An ammonium fluoride-ammonium nitrate solution was used in the PUREX plant during the 1983 to 1988 period to dissolve Zircaloy cladding from N Reactor fuel. The NCRW resulted when the spent cladding waste was made alkaline and stored in DSTs. The PFP waste resulted when composite acidic waste from the PFP was made alkaline and stored in a DST. The CC waste has a very high concentration of organic chelating agents and their degradation products. The CC waste is the concentrated aqueous raffinate from ^{90}Sr liquid-liquid extraction operations performed in the 1960's and 1970's. The DSS and DSSF wastes are viscous, highly-alkaline liquid wastes containing high concentrations of sodium salts generated from evaporation of dilute low level waste solutions. The DSS waste differs from DSSF waste in that it has been evaporated past the aluminate phase boundary and does not normally separate into sludge and supernatant liquids.

The solid portion of the NCAW, NCRW, CC waste, and PFP waste all contain > 100 nCi/g of TRU elements. The NCAW solids also contain the majority of the ^{90}Sr present in the PUREX process high-level radioactive waste; other DST solid wastes do not contain large concentrations of ^{90}Sr. The NCAW and CC alkaline waste solutions contain relatively high concentrations of ^{137}Cs. Because of the large amounts of organic chelators, some alkaline CC liquid waste also contains > 100 nCi/g of TRU elements.

Single-Shell Tank Wastes

Wastes in the SSTs consist mainly of two types of solids: sludge and salt cake. A small amount (2,300 m^3) of interstitial liquid is also present. Sludges consist principally of metal (e.g., iron, chromium, nickel, aluminum, etc.) oxides and hydroxides. These precipitated when the acidic liquid wastes from the bismuth phosphate (BiPO$_4$), reduction-oxidation (REDOX), and PUREX processes were made alkaline before routing to the SSTs. Salt cake is mainly composed of water-soluble sodium salts (e.g., NaNO$_3$, Na$_2$CO$_3$, NaNO$_2$, NaOH, etc.) that crystallized when the original highly alkaline liquid wastes were evaporated. Of the SST radionuclide inventory, over 99 percent of the uranium, plutonium, and other TRU elements, ^{90}Sr, and some of the ^{99}Tc are in the sludge, while the salt cake contains over 90 percent of the ^{137}Cs and the rest of the ^{99}Tc.

Twenty four of the SSTs contain so-called ferrocyanide wastes. These wastes were generated in the late 1950s when aged $BiPO_4$ process wastes were hydraulically retrieved and processed to recover large amounts of valuable uranium. Sodium ferrocyanide and nickel sulfate solutions were added to the aqueous raffinate from the Metal (Uranium) Recovery Process to precipitate nickel ferrocyanide at pH 9-10. The nickel ferrocyanide precipitate carried down with it almost all the ^{137}Cs in the waste. The supernatant liquid remaining after settling of the nickel ferrocyanide and hydrated metal oxides met all then existing requirements for disposal to the ground. Nickel ferrocyanide solids in the 24 SSTs are of particular interest and concern because of their potential, under certain conditions, of reacting vigorously with dried nitrate and/or nitrite salts.

EVOLUTION OF STRATEGIES FOR DISPOSAL OF HANFORD TANK WASTES

The earliest studies to define and develop an acceptable strategy for disposal of Hanford Site tank wastes began in 1972 (Table 1). The early and preliminary studies were documented in 1977 with issuance of the first technical report (ERDA, 1977) on selection and evaluation of technical alternatives for long-term management of Hanford Site tank wastes. Additional engineering studies, supplemented by on-going bench-scale tests of waste treatment and disposal form technology, were performed in the period 1979-1985. The first phase (1972-1988) of the efforts to develop a tank waste disposal strategy concluded in 1988 with the issuance of a Record of Decision on the Final Environmental Impact Statement, Disposal of Hanford Defense High-Level, Transuranic, and Tank Waste (DOE, 1988).

Table 1. Significant Events In Evolving Strategy for Disposal of Hanford Site Tank Waste: 1972-1988.

Date	Event	Reference
1972 - 1977	Preliminary engineering studies and evaluations; preliminary bench-scale experiments with simulated and actual tank wastes	
1977	First technical report on alternatives for long-term management of Hanford Site high-level radioactive waste	ERDA, 1977
1980	Follow-on reports on alternatives for disposal of Hanford Site high-level wastes	RHO, 1980a RHO, 1980b RHO, 1980c
1983	Definitive engineering study on disposal of DST wastes	Schulz et al., 1983
1985	Hanford Defense Waste Disposal Alternatives Engineering Support Data for the Hanford Defense Waste Environmental Impact Statement	RHO, 1985
1987	Final Environmental Impact Statement, Disposal of Hanford Defense High-Level, Transuranic, and Tank Wastes, Hanford Site, Richland, Washington	DOE, 1987
1988	Record of decision on final environmental impact statement	DOE, 1988

DOE = U.S. Department of Energy
DST = Double-shell tank

The 1988 Record of Decision stated that:

- The DST wastes would be retrieved and separated into two fractions, with the high-level fraction vitrified and disposed of in a deep geologic repository and the low-level fraction grouted in near-surface facilities. It was generally accepted that any pretreatment of DST waste to separate it into high- and low-level waste fractions would be done in the Hanford B Plant, a facility constructed during Manhattan Project days in the early 1940s.

- A final decision on disposal of SST wastes would not be made until additional development and evaluation studies were completed.

- Capsules of $^{90}SrF_2$ and $^{137}CsCl$ produced in the 1960-1985 time frame from selected retrieved wastes would be disposed of in a deep geologic repository.

Figure 3 illustrates the 1988 version of the Hanford Site plan for disposal of tank wastes and stored capsules. This plan envisioned that some SST wastes, perhaps most, would be left in place for disposal and that some SST waste might have to be retrieved. The 1988 disposal plan called for disposal of cesium and strontium capsules directly in a deep geologic repository after, perhaps, repackaging them in canisters suitable for geologic disposal.

Events (Table 2) which occurred in 1989 immediately after the Record of Decision was issued supported the 1988 tank waste disposal strategy. Thus, in 1989 the U.S. Department of Energy, U.S. Environmental Protection Agency, and the Washington State Department of Ecology co-signed the Hanford Federal Facility Agreement and Consent Order (Tri-Party Agreement)(Ecology et al., 1992). This Tri-Party Agreement established a strategy, schedule, and milestones for disposal of all the tank wastes. Also, engineering studies in 1989 and 1990 provided updated assessments of proposed DST pretreatment technologies and facilities.

Several recent events which occurred in 1991 combined to drastically alter the 1988 plans for disposal of tank wastes. For example, the Washington State Department of Ecology indicated in 1991 that the 40-year old B Plant did not meet minimum licensing and permitting regulations for extended use as a central facility for pretreating DST wastes. Also, in 1991, 6 DSTs and 48 SSTs, including the 24 SSTs containing ferrocyanide wastes, were found to have unreviewed safety questions requiring high priority mitigation and remediation. These events prompted a comprehensive engineering study to redefine the baseline plan for disposal of tank wastes; results of this study were published in late 1991.

Continued development of a new and completely satisfactory strategy for disposal of Hanford Site tank wastes was further complicated by DOE's decision in 1992 to proceed with all the legal steps required to make a decision to retrieve all SST wastes for pretreatment and disposal. At the same time, the DOE established the Tank Waste Remediation System (TWRS) to address safe interim management of both SST and DST wastes; mitigation and remediation of all tanks with unreviewed safety questions; and retrieval, pretreatment, and disposal of all tank wastes.

In response to the 1992 DOE actions, Westinghouse Hanford Company with assistance from Battelle's Pacific Northwest Laboratory conducted extensive and exhaustive examination of several technical options for disposal of Hanford Site tank waste. Results of these studies, published in 1993, were used by the DOE to establish the current TWRS strategy (Figure 4). This strategy embraces three key activities:

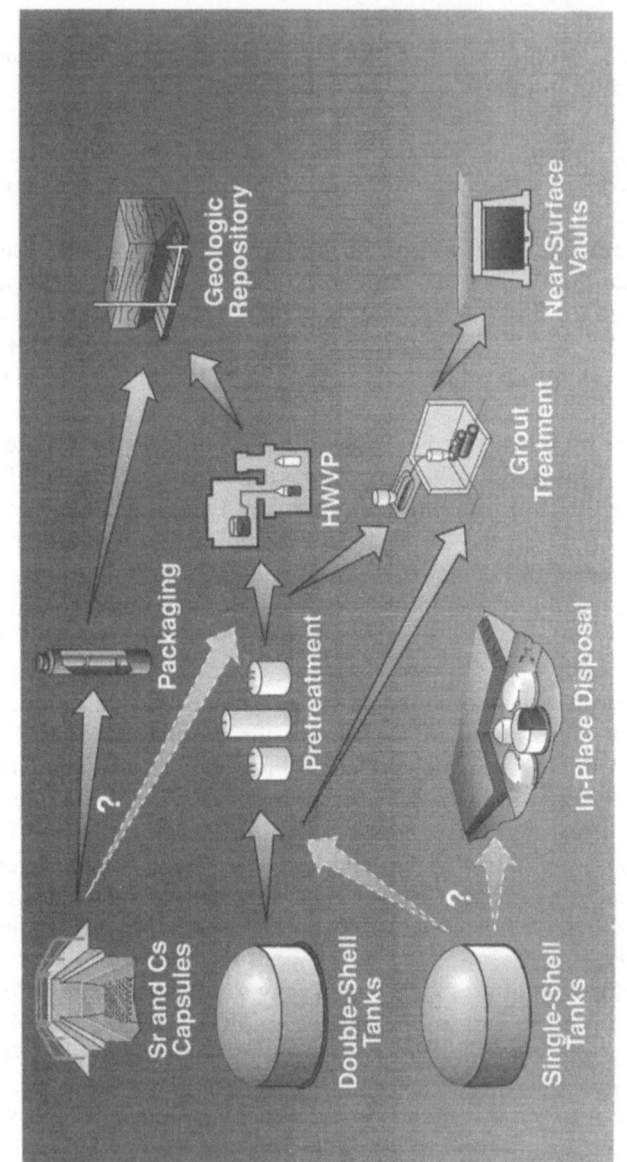

Figure 3. Hanford Site Radioactive Waste Tanks and Capsule Disposal Plan 1988-1991

Table 2. Significant Events in Evolving Strategy for Disposal of Hanford Tank Wastes: 1989-1993.

Date	Event	Reference
1989	Hanford Federal Facility Agreement and Consent Order signed by DOE, EPA, and Ecology	Ecology et al., 1990
1989	Updated assessment of processes and facilities for pretreating DST waste	Kupfer et al., 1989
1989	Further updated assessment of DST waste pretreatment alternatives	WHC, 1990
1991	Washington state Department of Ecology advises that B Plant does not meet minimum licensing and permitting criteria	Wodrich, 1992
1991	6 DSTs and 48 SSTs found to have unreviewed safety questions requiring high priority mitigation and remediation	Wodrich, 1992
1991	Tank Waste Disposal Program Redefinition	Grygiel et al., 1991
1992	DOE announces decision to proceed with NEPA documentation for retrieval of all SST wastes for pretreatment and disposal	
1992	Secretary of Energy provides initial set of goals and objectives for the newly-established Tank Waste Remediation System	Wodrich, 1992
1993	Tank Waste Technical Options report discusses technical details and projected costs for various options for disposal of tank wastes	Boomer et al., 1993
1993	Tank Waste Decision Analysis Report establishes current strategy for disposal of Hanford Site tank wastes	WHC, 1993a

DOE = U.S Department of Energy
Ecology = Washington State Department of Ecology
EPA = Environmental Protection Agency
NEPA = National Environmental Protction Agency

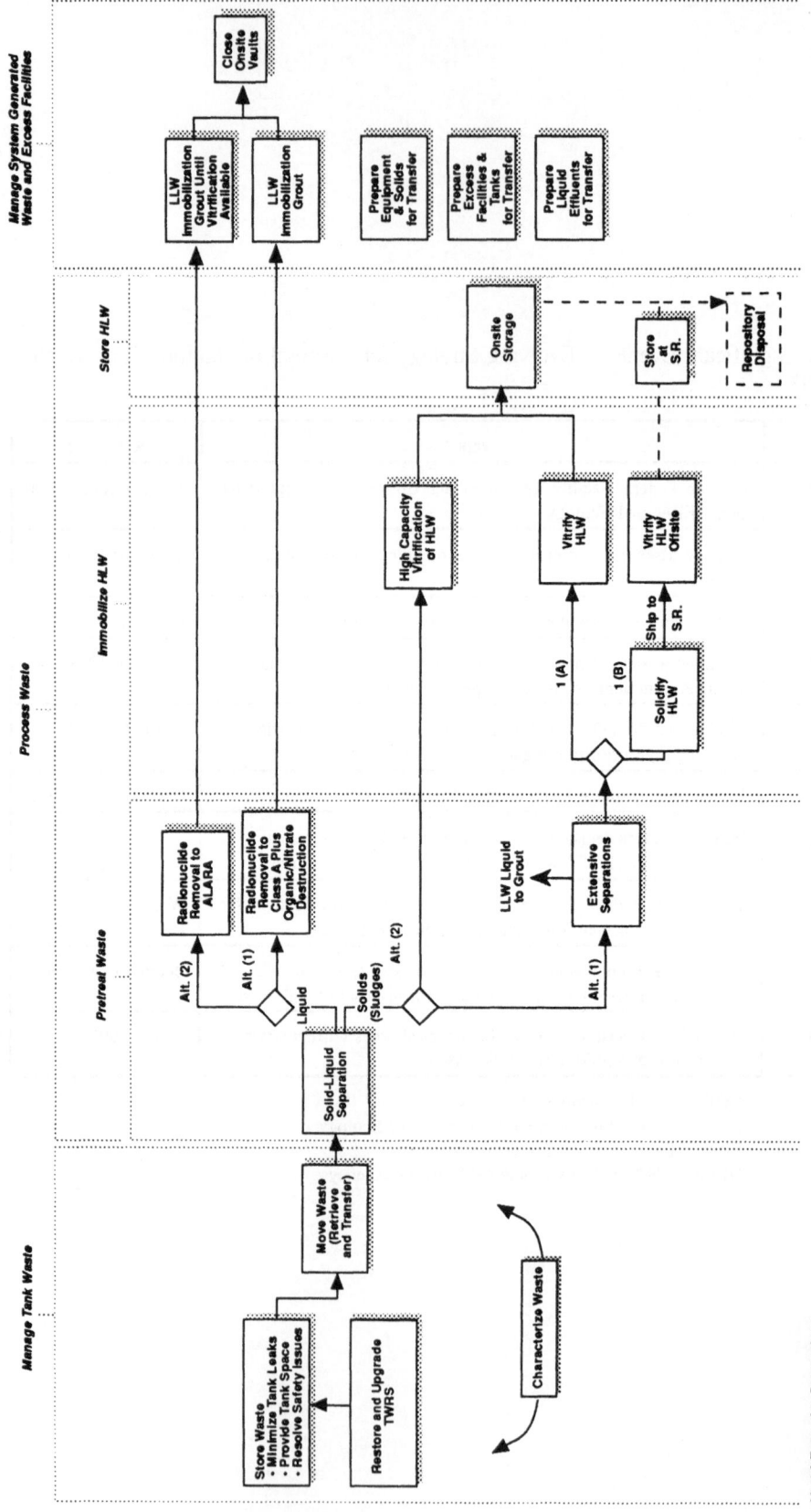

Figure 4. Tank Waste Remediation System Proposed New Technical Strategy.

Alt. (1) Extensive Separations
Alt. (2) High Capacity Vitrification

- Manage Tank Waste

- Process Tank Waste

- Manage System-Generated Waste and Excess Facilities

Manage Tank Waste

The proposed new TWRS strategy for managing tank wastes focuses first upon mitigation and/or remediation of tank safety issues. It also includes establishment of safety and environmental basis envelopes for continued tank farm operations. The tank farm infrastructure will be upgraded to compliance with today's standards.

Process Waste

The proposed new technical strategy for processing tank waste addresses changes that have occurred since the current baseline was developed. Early progress towards disposal will be initiated by retrieval of supernatant liquid and/or salt cake and subsequent processing to remove selected radionuclides. These radionuclide removal operations will be performed either in distributed compact processing units or in a to-be-constructed initial pretreatment module.

For processing sludge, two bounding technologies will be developed in parallel until sufficient technical data are available to permit decision makers to choose a disposal processing method. The bounding technologies are as follows:

- Extensive Separations. Technology, much of it still being developed, will be used to separate retrieved tank wastes into a small volume of high-level waste and a large volume of low-level radioactive waste. The goal is to produce low-level radioactive waste with greatly reduced chemical toxicity and whose radionuclide concentrations are within Nuclear Regulatory Commission limits for Class A waste. The small volume of high-level radioactive waste will be vitrified for eventual deep geologic disposal.

- High-Capacity Vitrification. Currently available and relatively simple technology will be used to separate retrieved tank wastes into high-level and low-level fractions. The high-level fraction will be vitrified, while the low-level radioactive waste fraction may be converted to a solid cementitious product for near-surface disposal. In this bounding technology, the low-level radioactive waste fraction would meet regulatory requirements for near-surface disposal but would contain higher radionuclide concentrations than the waste obtained in the extensive separations approach.

Manage System-Generated Waste and Excess Facilities

Low-level radioactive waste generated during processing of tank wastes will be disposed of as grout or some other compliant form. Other secondary wastes will be prepared for disposal to meet applicable regulatory requirements. Excess facilities will be transferred to the Environmental Restoration Program for decontamination and decommissioning.

PRETREATMENT TECHNOLOGIES

As noted in the previous section, pretreatment of Hanford Site tank wastes prior to their disposal has long been and continues to be a highly important element of all disposal strategies. Pretreatment processes are essentially partitioning processes to divide the tank wastes into a high-level radioactive waste fraction and a low-level radioactive waste fraction. Envisioned pretreatment processes range from simple operations to very complex chemical separations processes. Thus, pretreatment processes currently under consideration for application to Hanford Site tank wastes include the following:

- Water dissolution of salt cake

- Solid-liquid separations, e.g., salt cake solution from sludges, etc.

- Acid dissolution of sludges

- Removal of ^{137}Cs from salt cake and/or sludge solutions

- Removal of ^{99}Tc from salt cake and/or sludge solutions

- Removal of ^{90}Sr from sludge solutions and, as needed, salt cake solutions

- Removal of transuranium (TRU) elements from sludge solutions and, as needed, salt cake solutions

- Destruction of organic chelators in both alkaline and acidic waste solutions

The simplest pretreatment scheme involves treatment of the retrieved salt cake-sludge mixture with water to solubilize sodium salts and most of certain radionuclides, e.g., ^{137}Cs and ^{99}Tc. Appropriate solid-liquid separation techniques, e.g., decantation, filtration, etc., are used to separate the salt cake solution from the undissolved sludge. After washing with water, the separated sludge is vitrified. Selected radionuclides, e.g., ^{137}Cs, ^{99}Tc, and, perhaps, others are removed from the salt cake solution and the sludge washes before their incorporation and solidification in a grout form for onsite disposal. This simple pretreatment approach, first studied by Schulz (1980), would be applied if the High Capacity Vitrification bounding technology approach of the TWRS strategy is implemented.

Pretreatment of the washed sludge to reduce the mass of material, mainly nonradioactive constituents, which must be vitrified involves additional processes to dissolve it and remove selected radionuclides, e.g., TRU elements, ^{90}Sr, ^{99}Tc, and, possibly, ^{137}Cs. Potential reagents and procedures which may be applied to aqueous dissolution of various types of Hanford Site SST sludges have been discussed by Schulz and Kupfer (1991).

Various aqueous-based processes are available (Table 3) for removing selected individual radionuclides from acidic and, in some cases, alkaline media. A bounding process, the so-called CLEAN process (Straalsund 1992), has been proposed for removal of TRU elements, ^{90}Sr, ^{99}Tc, and ^{137}Cs from acid solutions of Hanford SST sludges. [The CLEAN process is an adaptation of the previously known CURE (Clean Use of Energy) concept (WHC, 1990)]. Two major objectives of the CLEAN process are (a) to generate a large volume of low-level radioactive waste meeting or exceeding Nuclear Regulatory Commission specifications for Class A waste and (b) to produce not more than 1,000 canisters (61 cm diameter x 3 m long) of vitrified high-level radioactive waste. To achieve this latter goal, the CLEAN process also entails processes for separation of trivalent actinide elements from trivalent lanthanides and of ^{90}Sr from any associated nonradioactive barium.

It is not yet clear whether the CLEAN or some less ambitious radionuclide removal scheme will be implemented on a plant-scale at the Hanford Site. What is clear is that some form of pretreatment of Hanford Site tank wastes continues to be show great advantages in schemes for final disposal of such wastes. It is also clear that, for some time in the future, scientists and engineers in various DOE laboratories and elsewhere will spend considerable energy in modifying, developing, and demonstrating pretreatment technology for use in disposal of Hanford Site SST and DST wastes.

Table 3. Listing of Representative Candidate Technologies for Pretreatment of Hanford Tank Wastes.

	Candidate Technology	Reference
1.	Sludge Dissolution	
	Aqueous Dissolution Procedures	Schulz and Kupfer, 1991
	Fusion Dissolution Procedures	Schulz and Kupfer, 1991
2.	TRU Element Removal[a]	
	TRUEX SX[b] Process	Horwitz and Schulz, 1985
	CMP SX Process	WHC, 1993b
	Diamide SX Process	Cuillerdier et al., 1993
3.	^{99}Tc Removal	
	TRUEX SX Process[a]	WHC, 1993b
	CMP SX Process[a]	WHC, 1993b
	Anion Exchange Resins[c]	Roberts et al., 1962
4.	^{90}Sr Removal	
	SREX SX Process[a]	Horwitz et al., 1991
	Inorganic Sorbents[a,d]	Schulz and Bray, 1987
	Cobalt Dicarbollide SX Process[a,c]	Reilly et al., 1992
5.	^{137}Cs Removal	
	Cobalt Dicarbollide SX Process[a,c]	Reilly et al., 1992
	Inorganic Sorbents[a,e]	Schulz and Bray, 1987
	Organic Sorbents[c,f]	Bibler and Wallace, 1987
	Precipitation[a,c,g]	Schulz and Bray, 1987
6.	Destruction of Organic Chelates	
	Ozonation[c]	Bollyky and Beary, 1981
	Calcination	Colby, 1993

(a) From strongly acidic solutions
(b) SX = Solvent Extraction
(c) From alkaline solutions
(d) For example, antimonic acid
(e) For example, zeolites, crystalline silicotitanates, etc.
(f) For example, formaldehyde- or resorcinol-formaldehyde-based cation exchange resins
(g) For example, cesium phosphotungstate, cesium tetraphenylborate, nickel ferrocyanide, etc.

REFERENCES

Bibler, J. P. and R. M. Wallace, 1987, Preparation and Properties of a Cesium-Specific Resorcinol-Formaldehyde Ion Exchange Resin, U.S. DOE Report DPST-87,647, E. I. duPont de Nemours, Co., Aiken, South Carolina.

Bollyky, J. L. and M. M. Beary, 1981, Ozone Mass Transfer and Kinetics, RHO-C-47, Rockwell Hanford Operations, Richland, Washington.

Boomer, K. D., 1993, Tank Waste Technical Options Report, WHC-EP-0616 Westinghouse Hanford Company, Richland, Washington.

Colby, S. A., 1993, Three Candidate Treatment Technologies for Remediation of Hanford Site Watch List Tanks, WHC-SA-1825-FP, Westinghouse Hanford Company, Richland, Washington.

Cuillerdier, C., C. Musikas, and L. Nigond, 1993, Sep. Sci. & Tech., 28:1

DOE, 1987, Final Environmental Impact Statement, Disposal of Hanford Defense High-Level and Transuranic Wastes, Hanford Site, Richland, Washington, DOE/EIS-0113, Vol. 1 through 5, U.S. Department of Energy, Richland, Washington.

DOE, 1988, Final Environmental Impact Statement for the Disposal of Hanford Defense High-Level, Transuranic, and Tank Wastes, Hanford Site, Richland, Washington: Record of Decision, "Federal Register," Vol. 53, No. 72, pp. 12449-12453, U.S. Department of Energy, Washington, D.C.

Ecology, EPA, and DOE, 1992, Hanford Federal Facility Agreement and Consent Order, 2 vols., as amended, Washington State Department of Ecology, U.S. Environmental Protection Agency, and U. S. Department of Energy, Washington, D.C.

ERDA, 1977, Alternatives for Long-Term Management of Defense High-Level Radioactive Waste, ERDA-77-44, U. S. Energy Research and Development Administration, Richland, Washington.

Grygeil, M. L., 1991, Tank Waste Disposal Program Redefinition, WHC-EP-0475 Westinghouse Hanford Company, Richland, Washington.

Horwitz, E. P. and W. W. Schulz, 1985, "Solvent Extraction and Recovery of Transuranium Elements from Waste Solutions Using the TRUEX Process," in Solvent Extraction and Ion Exchange in the Nuclear Fuel Cycle, Ellis Horwood, Ltd., Chichester, England, p. 137.

Horwitz, E. P., M. L. Dietz, and D. E. Fisher, 1990, Solvent Extr. Ion. Exch., 8:199

Kupfer, M. J., A. L. Boldt, and J. L. Buelt, 1989, Process and Facility Options for Pretreatment of Hanford Tank Waste, SD-WM-TA-015, Westinghouse Hanford Company, Richland, Washington.

Reilly, S. D., C. F. V. Mason, and P. H. Smith, 1992, Cobalt(III) Dicarbollide, A Potential ^{137}Cs and ^{90}Sr Waste Extraction Reagent, U. S. DOE Report LA-11695, Los Alamos National Laboratory, Los Alamos, New Mexico .

RHO, 1980a, Technical Status Report on Environmental Aspects of Long-Term Management of High-Level Defense Waste at the Hanford Site, RHO-LD-139, Rockwell Hanford Operations, Richland, Washington.

RHO-1980b, Environmental Aspects of Long-Term Management Alternatives for High-Level Defense Waste at the Hanford Site, RHO-LD-140, Rockwell Hanford Operations, Richland, Washington.

RHO-1980c, Technical Aspects of Long-Term Management Alternatives for High-Level Defense Waste at the Hanford Site, RHO-LD-141, Rockwell Hanford Operations, Richland, Washington.

RHO, 1985, <u>Hanford Defense Waste Disposal Alternatives: Engineering Support Data for the Hanford Defense Waste-Environmental Impact Statement</u>, RHO-RE-ST-30P, Rockwell Hanford Operations, Richland, Washington.

Roberts, F. P., Smith, F. M., and Wheelwright, E. J., 1962, <u>Recovery and Purification of Technetium-99 from Neutralized PUREX Wastes</u>, U. S. AEC Report, General Electric Company, Richland, Washington.

Schulz, W. W., 1980, <u>Removal of Radionuclides from Hanford Defense Waste Solutions</u>, RHO-SA-51, Rockwell Hanford Operations, Richland, Washington.

Schulz, W. W., M. J. Kupfer, and J. P. Sloughter, 1983, <u>Evaluation of Process and Facility Options for Treatment of Double-Shell Tank Wastes</u>, SD-WM-ES-023, Rockwell Hanford Operations, Richland, Washington.

Schulz, W. W. and L. A. Bray 1987, <u>Sep. Sci. & Tech.</u>, 22:191.

Schulz, W. W. and M. J. Kupfer, 1991, <u>Candidate Reagents and Procedures for the Dissolution of Hanford Site Single-Shell Tank Sludges</u>, WHC-EP-0451, Westinghouse Hanford Company, Richland, Washington.

Straalsund, J. L., et al., 1992, <u>Clean Option: An Alternative Strategy for Hanford Tank Waste Remediation</u>, PNL-8388, Vol. 1, Pacific Northwest Laboratory, Richland, Washington.

WHC, 1990, <u>Assessment of Double-Shell Tank Waste Pretreatment Options</u>, WHC-SP-0464, Westinghouse Hanford Company, Richland, Washington .

WHC, 1990, <u>CURE: Clean Use of Reactor Energy</u>, WHC-EP-0268 Westinghouse Hanford Company, Richland, Washington.

WHC, 1993a, <u>Tank Waste Decision Analysis Report</u>, WHC-EP-0617, Westinghouse Hanford Company, Richland, Washington.

WHC, 1993b, <u>Alternative Pretreatment Technologies for Removal of Transuranium Elements from Selected Hanford Site Wastes</u>, WHC-EP-0577 Westinghouse Hanford Company, Richland, Washington.

Wodrich, D. D., 1992, <u>Hanford Site Tank Waste Remediation System</u>, WHC-SA-1545-FP, Westinghouse Hanford Company, Richland, Washington .

PROCESS CHEMISTRY FOR THE PRETREATMENT
OF HANFORD TANK WASTES

Gregg J. Lumetta[1], John L. Swanson[1], and Steven A. Barker[2]

[1]Pacific Northwest Laboratory[a]
Richland, Washington 99352
[2]Westinghouse Hanford Company
Richland, Washington 99352

INTRODUCTION

Methods are being developed to treat and dispose of large volumes of radioactive wastes stored in underground tanks at the U.S. Department of Energy's (DOE) Hanford Site. Under current guidelines, the high-level waste (HLW) will be vitrified into borosilicate glass and disposed of in a geologic repository, and the low-level waste (LLW) will be converted to grout and disposed of by shallow land burial on the Hanford Site. Because of the high cost of vitrification and geologic disposal, pretreatment methods are being developed to minimize the volume of HLW requiring disposal.

In this paper, we discuss two general approaches that are being considered for pretreating Hanford tank sludges. In the first approach, critical components are leached from the sludges, leaving the transuranium (TRU) elements in the sludge. In the second approach, acid dissolution of the sludges is followed by separating the TRU elements from the bulk sludge compentents using the transuranium extraction (TRUEX) process[1,2,3,b].

Vitrification of Hanford Site HLW will be performed in a liquid-fed ceramic melter, but certain elements in the vitrification feed cause problems in operating such a melter. For example, chromium can form an undesirable crystalline phase that can cause problems in operating the vitrification plant. Thus, the limits for chromium in the vitrification feed are low.

For any given waste, a critical component is present that defines the minimum number of glass canisters required to dispose of that waste. In the case of Hanford

---

[a]Pacific Northwest Laboratory is operated for the U.S. Department of Energy by Battelle Memorial Institute under Contract DE-AC06-76RLO 1830.
[b]Depending on the composition of the waste and the specifications for the LLW form, additional radionuclides (e.g., ^{90}Sr) might also need to be removed.

Chemical Pretreatment of Nuclear Waste for Disposal, Edited by
W.W. Schulz and E.P. Horwitz, Plenum Press, New York, 1995

Plutonium Finishing Plant (PFP) sludge,[a] the critical component is chromium. The next critical component is phosphorus; that is, if the chromium were removed from this sludge, phosphorus would become the limiting component. The third and fourth critical components are sulfur and aluminum, respectively. Removing the critical component results in lowering the number of glass canisters required to dispose of the waste (Table 1). Methods to leach these four components from PFP sludge are described in this paper.

Table 1. Effect of the removal of critical components from PFP sludge on the number of glass canisters required to dispose of this waste.

Option	No. of Canisters Needed	Limiting Component
Sludge Wash (SW) Only	2480	Cr
SW+Cr removal	1230	P
SW+Cr+P removal	750	S
SW+Cr+P+S removal	680	Al
SW+Cr+P+S+Al removal	300	Other

An alternative approach to the pretreatment of Hanford tank sludges is to dissolve the sludge in acid and separate the relatively small amount of TRU elements from the bulk sludge components.[b] The TRUEX process is being developed for this purpose[1,2,3]. This process is a solvent extraction process in which the TRUs are extracted from nitric acid solution using octyl(phenyl)-N,N-diisobutylcarbamoylmethylphosphine oxide (CMPO). In this paper, we discuss the TRUEX processing of another unique Hanford waste—neutralized cladding removal waste (NCRW) sludge.[a]

EXPERIMENTAL

Leaching of PFP Sludge

A 0.65-g sample of PFP sludge was mixed with 5 mL of 0.1 $\underline{M}$ NaOH, and the slurry was heated at 100°C for 1 h. After cooling, the mixture was centrifuged, and the supernatant liquid was decanted. This treatment with 0.1 $\underline{M}$ NaOH was repeated three more times.

---

[a]The PFP sludge consists of the neutralized raffinate from a tributyl phosphate (TBP) extraction process for recovering Pu from scrap materials. This process was conducted in the PFP at Hanford. Miscellaneous other wastes have also been added to the tank containing this sludge. The sludge consists primarily of Fe, Al, Cr, Mn, and Ca oxides/hydroxides contaminated with transuranic elements and various fission products.

[b]NCRW sludge was formed by the neutralization of the solution formed by chemical decladding of Zircaloy-clad metallic uranium fuel by the Zirflex Process[4]. The sludge consists primarily of $Zr(OH)_4$ and NaF. It also contains U, TRUs (Am + Pu ~1000 nCi/g sludge), and mixed fission products (^{137}Cs, ^{90}Sr, ^{125}Sb, etc.).

The PFP sludge was then mixed with 10 mL of 0.014 $\underline{M}$ KMnO$_4$ in 0.1 $\underline{M}$ NaOH at 100°C for 1 h. After cooling, the mixture was centrifuged, and the leach solution was decanted. The leached sludge was dissolved for analysis by heating at 100°C with 13.5 mL of 1.8 $\underline{M}$ HCl/1.1 $\underline{M}$ HF solution.

All of the solutions were analyzed for aluminum, chromium, iron, and phosphorus by inductively coupled plasma atomic emission spectroscopy (ICP). Ion chromatographic analyses were done to determine the PO$_4^{3-}$ and SO$_4^{2-}$ content of each solution.

TRUEX Processing of NCRW Sludge

A 2.45-g portion of NCRW sludge was slurried with 19 mL of water. Nitric acid (3.8 mL of 15.7 $\underline{M}$) was added dropwise with stirring. After the last addition of HNO$_3$, the mixture was stirred for 2 h at room temperature. The dissolved NCRW sludge solution was clarified by filtration through a 0.2-μm membrane filter.

To prepare the feed for the solvent extraction process, 8.33 mL of water and 0.63 mL of 15.7 $\underline{M}$ HNO$_3$ were added to 15.0 mL of the dissolved NCRW sludge solution. After 0.35 mL of this solution was removed for various analyses, 0.24 mL of 1 $\underline{M}$ H$_2$C$_2$O$_4$ was added to complete preparation of the aqueous phase for the first extraction contact. This solution represented a blend of the feed solution and the three scrub streams shown in Figure 1.

The TRUEX process solvent consisted of 0.2 $\underline{M}$ CMPO plus 1.4 $\underline{M}$ TBP dissolved in a mixture of normal paraffin hydrocarbons (NPH). It was washed with aqueous carbonate solution prior to use.

Eight milliliters of the TRUEX process solvent were mixed with 24 mL of the feed solution for 30 sec. The mixture was centrifuged to ensure separation of the two phases. Portions of both the organic and aqueous phases were subjected to additional contacts as summarized in Table 2.

The concentrations of nonradioactive metal ions were determined by ICP analysis. Analyses for alpha-emitting radionuclides involved counting 0.1-mL dried mounts of diluted samples from the aqueous phase of each contact. Concentrations of the various components in the organic phases were calculated by mass balance. Total fluoride concentrations were determined potentiometrically using a fluoride-selective electrode, and acid content was determined by potentiometric titration with standard NaOH. The acid concentrations reported in this paper represent the concentration of HNO$_3$ + HF. The endpoints of the titrations were taken to be at pH 7, and the acid concentrations were corrected for the contribution of hydrolyzable ions using the following equation: $[H^+] = [H^+]_{total} - 2[Zr] - 3([Al] + [Fe] + [Cr])$.

RESULTS AND DISCUSSION

Leaching of PFP Sludge

When a portion of PFP sludge was digested with 0.1 $\underline{M}$ NaOH at 100°C, virtually all of the PO$_4^{3-}$ and SO$_4^{2-}$, and most of the aluminum, was removed from the sludge (Table 3). A fraction (27.1%) of the chromium originally present in the sludge was also removed by this treatment. This portion of the chromium was present as CrO$_4^{2-}$ (as determined spectrophotometrically), so it was readily soluble in 0.1 $\underline{M}$ NaOH. The remaining chromium in the sludge was in the +3 oxidation state; thus, to leach the remaining chromium from the sludge, it was oxidized to CrO$_4^{2-}$ with KMnO$_4$.

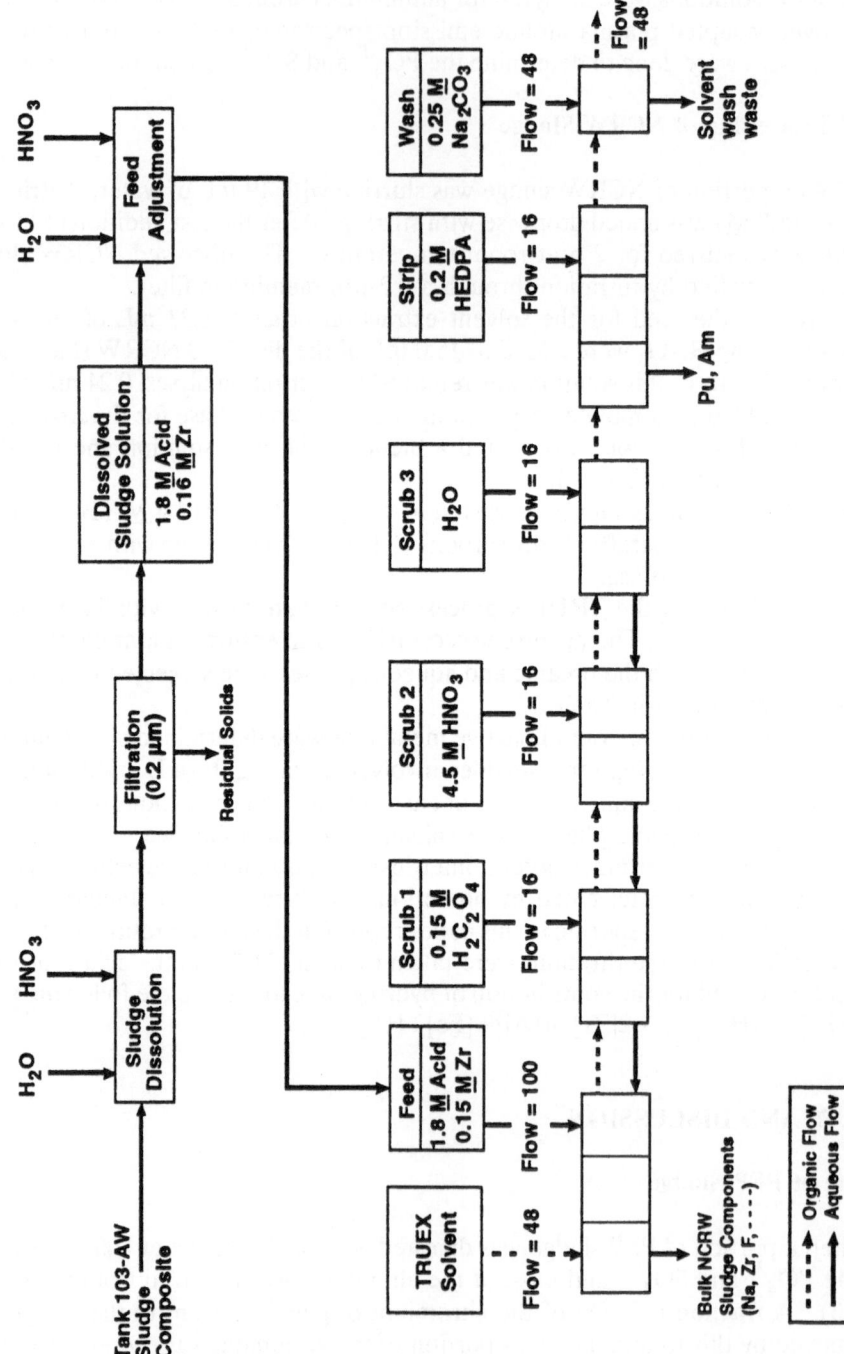

Figure 1. Flowsheet for the pretreatment of NCRW sludge by the TRUEX process.

Table 2. Summary of TRUEX process solvent extraction contacts performed to test the conceptual flowsheet depicted in Figure 1.

Contact	Step	Aqueous Phase	Organic Phase	O/A^a	Organic Vol., mL
A	Extn 1	Feed^b	TRUEX Process Solvent	0.33	8.00
B	Extn 2	From A	TRUEX Process Solvent	0.33	1.00
C	Extn 3	From B	TRUEX Process Solvent	0.33	0.67
D	Scrub 1	1.5 $\underline{M}$ HNO$_3$ + 0.05 $\underline{M}$ H$_2$C$_2$O$_4$^{c}	From A	1.00	7.00
E	Scrub 2	1.5 $\underline{M}$ HNO$_3$ + 0.05 $\underline{M}$ H$_2$C$_2$O$_4$	From D	1.00	6.00
F	Scrub 3	2.3 $\underline{M}$ HNO$_3$^{d}	From E	1.50	5.00
G	Scrub 4	2.3 $\underline{M}$ HNO$_3$	From F	1.50	4.00
H	Scrub 5	Water	From G	3.00	3.50
I	Scrub 6	Water	From H	3.00	3.00
J	Strip 1	0.21 $\underline{M}$ HEDPA^e	From I	3.00	2.50
K	Strip 2	0.21 $\underline{M}$ HEDPA^e	From J	3.00	2.00
L	Strip 3	0.21 $\underline{M}$ HEDPA^e	From K	3.00	1.50
M	Wash	0.25 $\underline{M}$ Na$_2$CO$_3$	From L	1.00	1.00

^{a}O/A = Volume of the organic phase divided by the volume of the aqueous phase.
^{b}The feed solution was a diluted NCRW sludge solution (see text).
^{c}Composition represents a blend of the three scrub streams shown in Figure 1.
^{d}Composition represents a blend of the second and third scrub streams shown in Figure 1.
^{e}HEDPA = 1-Hydroxyethane-1,1-diphosphonic acid.

Table 3. Results from the leaching of PFP sludge.

	Al	Cr	P	S	Fe
% in NaOH wash	69.1	27.1	100	100	0.2
% in KMnO$_4$ leach	29.8	69.3	0	0	0
% in leached sludge	1.1	3.6	0	0	99.8

The oxidation of Cr(III) to Cr(VI) with MnO$_4^-$ proceeds in basic solution according to the following equation:

$$Cr(OH)_3 + MnO_4^- + OH^- \rightleftharpoons CrO_4^{2-} + MnO_2 + 2H_2O \qquad \epsilon^0 = +0.72 \text{ V} \quad \textbf{(1)}$$

For every mole of chromium leached from the sludge, one mole of manganese is added in the form of MnO$_2$. Adding manganese to the sludge is acceptable because the limit for manganese in the vitrification feed is ten times the limit for chromium. However, by using MnO$_4^-$ to leach chromium, manganese becomes the critical component in the waste. It is estimated that if aluminum, phosphorus, and sulfur were removed by digestion in dilute NaOH, and chromium were removed by leaching with MnO_4^-, 420 canisters of glass would be required to dispose of the PFP sludge. This number could be lowered to approximately 300 canisters (Table 1) if a different oxidant were used. We are currently evaluating alternative oxidants for removing chromium.

If the PFP sludge is pretreated using the TRUEX process, it is expected that the number of glass canisters required to dispose of the TRU element fraction of the waste will be defined by the amount of the sludge that does not dissolve in HNO_3. Dissolution tests conducted in our laboratory suggest that approximately 100 canisters would be required if the PFP sludge were dissolved and processed by the TRUEX process. Thus, the TRUEX process approach would result in a smaller number of glass canisters being produced from PFP sludge than would the leaching approach. However, the TRUEX process would be much more difficult to implement than the simple leaching methods described here.

TRUEX Processing of NCRW Sludge

An early conceptual flowsheet for processing NCRW sludge by the TRUEX process is shown in Figure 1. Experiments performed to test this flowsheet involved the following steps: 1) sludge dissolution and solution clarification by filtration, 2) blending of the solution with the three scrub streams shown in Figure 1, 3) extraction, 4) scrubbing (with three different streams), and 5) stripping. The functions of the three scrub streams were to 1) remove extracted zirconium and fluoride with aqueous oxalic acid, 2) remove extracted oxalic acid with HNO_3, and 3) remove extracted HNO_3 with water.

Unwashed NCRW sludge dissolves readily in dilute HNO_3. In the treatment with HNO_3, 97.8% of the zirconium and virtually 100% of the sodium dissolved. Significant amounts of aluminum (25%) and phosphorus (43%) did not dissolve in HNO_3. The adjusted feed solution composition was 0.10 $\underline{M}$ Zr, 0.71 $\underline{M}$ Na, 0.033 $\underline{M}$ K, 0.024 $\underline{M}$ Al, 0.010 $\underline{M}$ Si, 0.004 $\underline{M}$ U, 0.37 $\underline{M}$ F, and 1.68 $\underline{M}$ H$^+$. The feed solution also contained 62.0 nCi/mL Pu and 9.5 nCi/mL Am.

The results of the extraction contacts are shown in Figure 2. The TRU elements were effectively extracted from the dissolved NCRW solution. The distribution coefficients for the TRU elements in the first and second extractions were 69 and 34, respectively; and approximately six in the third extraction contact. The raffinate from the third extraction contained only 0.1% of the TRU element activity originally in the feed. A significant amount of zirconium was extracted, but sodium and aluminum were not. As expected from published data,[1] a large fraction of the HNO_3 was extracted.

The small portion of zirconium that was extracted was removed from the solvent during the first three scrub contacts (Figure 3). Zirconium could not be detected in solutions from subsequent contacts. Thus, oxalic acid is very effective at removing zirconium from the extact. Of the fluoride that was extracted, 99.3% was scrubbed out after the sixth scrub contact; the data in Figure 3 suggest that the residual fluoride may be difficult to remove from the organic extract. The results from scrubs five and six indicate that water can lower the acid content of the extract. Throughout the six scrub contacts, the TRU elements remained in the organic phase.

The TRU elements were efficiently stripped from the scrubbed extract (Figure 4) with 1-hydroxyethane-1, 1-diphosphonic acid (HEDPA). Over 99% of the TRU elements were stripped in the first strip contact. The TRU elements remaining after the third strip were below detection limits.

These results serve as proof-of-principle that the TRUEX process can be used to separate TRU elements from the bulk components (zirconium and sodium) of NCRW sludge. Extraction of the TRU elements was very good; only ~0.1 nCi/mL of alpha activity remained in the aqueous phase after the third extraction. Good separation of the alpha-emitting radionuclides from most nonradioactive materials was achieved. For example, the feed solution contained ~7800 nCi TRU elements/g of zirconium, but the aqueous phase from the third extraction contained only 11 nCi/g zirconium. Likewise, the feed solution contained ~4400 nCi TRU elements/g of sodium, whereas the aqueous phase from the third extraction contained only 5 nCi TRU elements/g sodium. Elements that were not well separated from the TRU elements included uranium and the lanthanides.

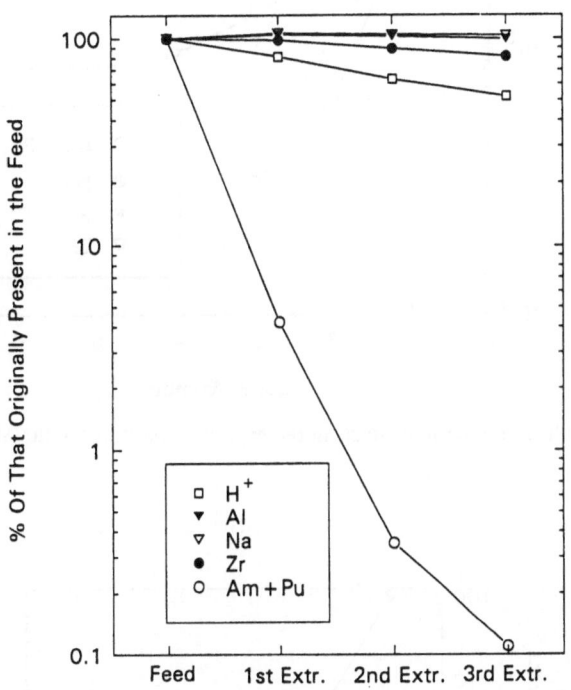

Figure 2 Results of the extraction contacts in the experiment testing the flowsheet in Figure 1.

Further studies identified some problems with processing NCRW sludge by the TRUEX process, such as

- corrosion of stainless steel piping and equipment
- precipitation from dissolved NCRW sludge solutions
- interfacial crud formation in the extraction step
- the need for excessive quantity of HEDPA stripping agent.

Each of these problems has been addressed and solved.

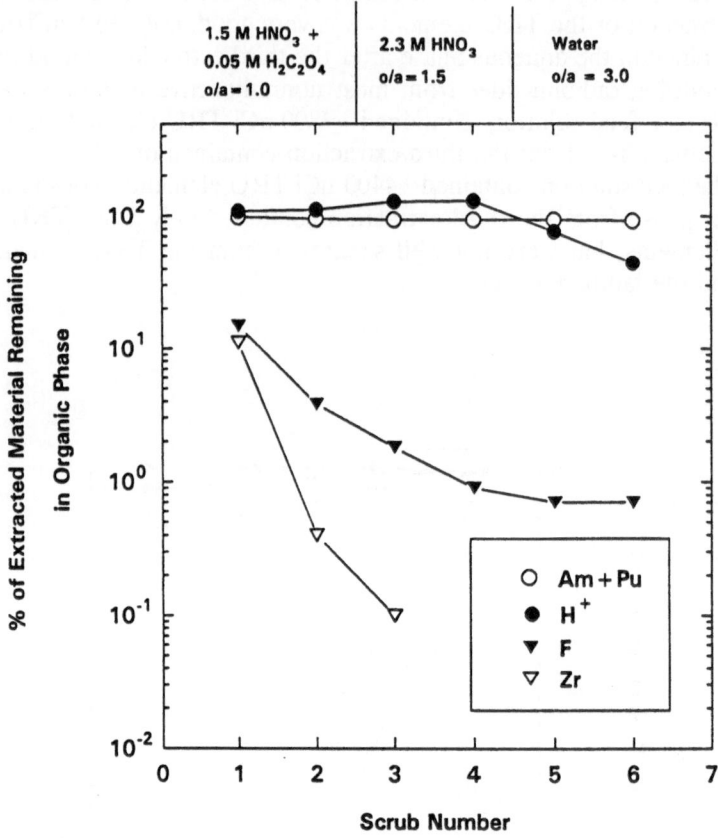

Figure 3 Results of the scrub contacts in the experiment testing the flowsheet in Figure 1.

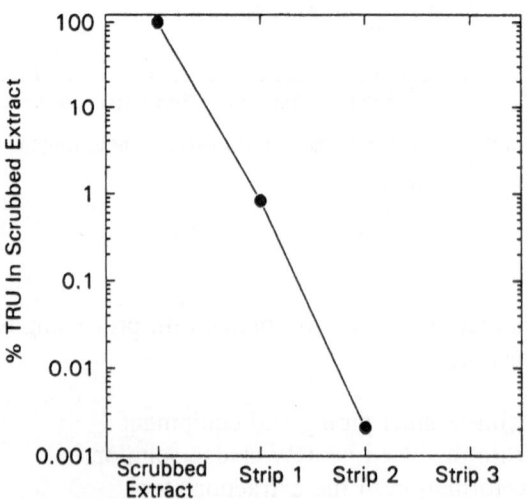

Figure 4 Results of the stripping contacts in the experiment testing the flowsheet in Figure 1.

Originally, it was planned to implement the TRUEX process in an existing facility (B-Plant) at the Hanford Site. Corrosion tests with simulated dissolved NCRW solutions indicated that the imbedded piping in this plant might be susceptible to corrosion while processing NCRW sludge by TRUEX. Recently, it has been determined that B-Plant will not be used to process Hanford Site tank wastes; rather, a new pretreatment facility will be built. Thus, corrosion of piping in B-Plant become a non-problem. Corrosion should not be a problem in the new facility because appropriate alloys will be chosen for construction of relevant equipment.

During the course of many experiments with actual NCRW samples, it was discovered that solids formed in dissolved sludge solutions at sporadic intervals ranging from weeks to days. The formation of precipitates in the dissolved sludge solution was deemed to be a problem because additional equipment clean-out capability would be required.

In these early experiments, the NCRW sludge samples were dissolved in HNO_3 directly. It was discovered that washing the sludge with water prior to acid dissolution resulted in dissolved sludge solutions that were stable for weeks to months (Table 4). Washing the sludge lowers the sodium and fluoride content. To achieve adequate sludge dissolution, fluoride must be added during the dissolution step so that the $F/(Zr+Al)$ ratio is approximately 2. This can be done by adding HF along with HNO_3 during dissolution. Solutions prepared by dissolving washed NCRW sludge in this manner are sufficiently stable that equipment clean-out capability is not expected to be necessary.

Table 4. Stability of dissolved sludge solutions toward precipitation.

Zr, M	Al, M	Na, M	H^+, M	$F/(Zr+Al)$	Days Stable
Unwashed Sludge:					
0.24	0.053	1.78	0.6	2.9	>1,<2
0.10	0.023	0.76	1.4	2.9	>1,<2
0.13	0.015	1.75	1.2	2.0	>20
Washed Sludge:					
0.18	0.006	0.20	1.7	1.8	>170
0.16	0.002	0.07	1.7	2.3	>5,<34
0.13	0.011	0.06	1.8	2.5	>225
0.12	0.001	0.05	1.5	1.3	>150

The third problem associated with processing NCRW sludge by the TRUEX process was interfacial crud which formed during the extraction step. In some cases, the amount of crud was so severe that the entire organic phase was filled with solids.

Further studies revealed that interfacial crud can be controlled by adjusting the $F/(Zr+Al)$ ratio in the solvent extraction feed. Figure 5 summarizes the interfacial crud observations from a number of experiments with dissolved NCRW solutions. Solutions were prepared from both washed and unwashed sludges. In all cases, the feed solutions were aged a minimum of 1 h before contacting the TRUEX process solvent; the organic-to-aqueous phase ratio was 0.33 in each contact. Although the process conditions have not yet been fully optimized, the data in Figure 5 suggest that

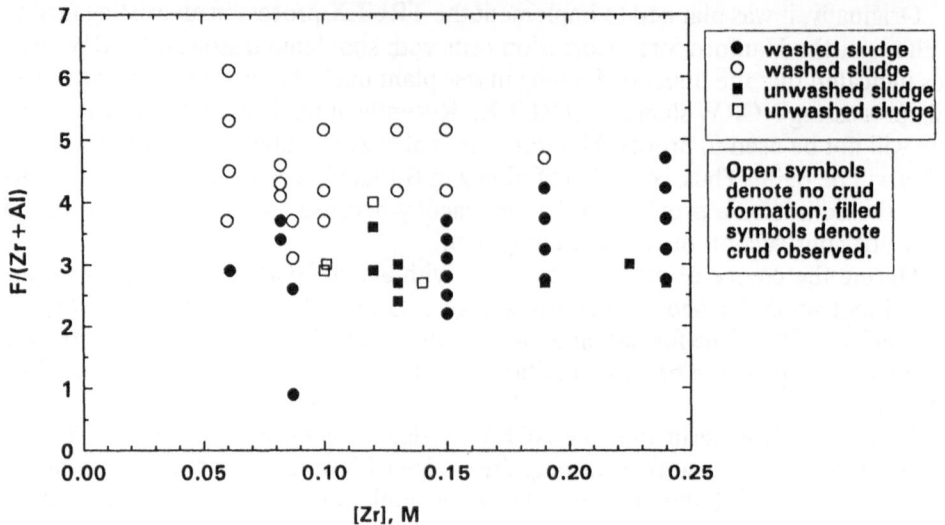

Figure 5. Interfacial crud formation in TRUEX process contacts with dissolved NCRW sludge. The F/(Zr+Al) ratio was adjusted by adding HF to the aqueous solutions. A minimum of 1 h was allowed before contacting with the TRUEX process solvent.

the concentration of zirconium in the extraction stages should be maintained below 0.15 $\underline{M}$ and the F/(Zr+Al) ratio greater than 4. The F/(Zr+Al) ratio can be adjusted by adding HF to the dissolved sludge solution. We have observed that increasing the F/(Zr+Al) ratio does not have an immediate effect in eliminating interfacial crud, so a lag time of 1 h should be allowed between adding HF and beginning the extraction step.

The composition of the interfacial crud has not been determined, but it is presumed to contain a zirconium species. The crud forms when the concentration of this species in the organic phase exceeds its solubility. Complexation of zirconium by fluoride lowers the distribution coefficient for zirconium; thus, raising the F/(Zr+Al) decreases the extraction of zirconium and the crud does not form.

The fourth problem with processing NCRW sludge by TRUEX concerns use of 0.2 $\underline{M}$ HEDPA as the stripping solution.[5] Although this solution performs very well in stripping the TRU elements from the extract (Figure 4), the phosphorus content of this solution will cause problems in subsequent vitrification of the TRU element fraction of the waste. Several alternative stripping solutions have been identified (e.g. sodium citrate, 0.002 $\underline{M}$ HEDPA + 0.25 $\underline{M}$ Na$_2$CO$_3$ [5], ferrous sulfamate, etc.). A final decision has not been made on which strip solution to use.

CONCLUSION

Two different approaches are being investigated for reducing the quantity of Hanford Site HLW needing disposal. Selective leaching of glass-limiting components is a method applicable in selected cases and is easy to implement. Simple methods have been demonstrated for leaching chromium, phosphorous, sulfur, and aluminum from PFP sludge. The second approach--sludge dissolution and extraction of TRU elements--is generally applicable and will likely result in a greater reduction in quantity of HLW needing disposal. Batch testing of the TRUEX process on actual NCRW sludge indicates that this process will effectively separate the TRU elements in this waste from the bulk sludge materials.

ACKNOWLEDGEMENTS

Pacific Northwest Laboratory is operated for the U.S. Department of Energy by Battelle Memorial Institute under Contract DE-AC06-76RLO 1830. The authors would like to thank Jeff Deal and Michael Wagner for performing much of the experimental work described in this paper. The assistance of Brian Rapko and Wayne Cosby in reviewing this paper is also gratefully acknowledged.

REFERENCES

1. E.P. Horwitz, D.G. Kalina, H. Diamond, G.F. Vandegrift, and W.W. Schulz, The TRUEX process - a process for the extraction of the transuranic elements from nitric acid wastes utilizing modified PUREX solvent, *Solvent Extr. Ion Exch.* 3:75-109 (1985).

2. E.P. Horwitz and W.W. Schulz, *Solvent extraction and recovery of the transuranic elements from waste solutions using the TRUEX process in: Solvent Extraction and Ion Exchange in the Nuclear Fuel Cycle*; D.H. Logsdail and A.L. Mills, eds., Ellis Horwood, Chichester pp. 137-144 (1985).

3. W.W. Schulz and E.P. Horwitz, The TRUEX process and the management of liquid TRU waste, *Sep. Sci. Technol.* 23:1191-1210 (1988).

4. J.L. Swanson, *The Zirflex process: in Progress in Nuclear Energy Series III: Process Chemistry*, F.R. Bruce, J.M. Fletcher, and H.H. Hyman eds., Pergamon Press, New York p. 289 (1961).

5. G.J. Lumetta and J.L. Swanson, Evaluation of 1-hydroxyethane-1, 1-diphosphonic acid and sodium carbonate as stripping agents for the removal of Am(III) and Pu(IV) from TRUEX process solvent, *Sep. Sci. Technol.* 28:43-58 (1993).

REMOVAL OF ACTINIDES FROM HANFORD SITE WASTES USING AN EXTRACTION CHROMATOGRAPHIC RESIN

G. S. Barney[1] and R. G. Cowan[2]

[1]Process Laboratories and Technology
[2]Process Development
Westinghouse Hanford Company
Richland, Washington 99352

INTRODUCTION

Economic incentives exist for partitioning both current and stored Hanford Site radioactive wastes into high-level waste, transuranic (TRU) waste, and low-level waste fractions. The present large volume of high-level and TRU wastes could be made much smaller by separating radionuclides from the inert (non-radioactive) components of the waste. Theoretically, the volume of high-level and TRU wastes could be reduced to only a few percent of the present volume. The remaining large volume of decontaminated, low-level waste could then be disposed of relatively inexpensively by mixing it with cement grout and disposing of the mixture in shallow, underground vaults. The smaller volume of high-level and TRU wastes could be disposed of by vitrification and transporting the vitrified waste to an underground waste repository. Recovery of some of the more valuable actinides in a relatively pure form for commercial use may also be an incentive for partitioning.

Several new technologies are being evaluated by Westinghouse Hanford Company for partitioning Hanford wastes. These technologies include the TRUEX process (Schulz and Horwitz, 1988) for separating TRU elements from acidic wastes. This solvent extraction process converts the bulk of TRU wastes into low-level wastes that can be disposed of inexpensively. Extraction chromatographic resins combine the selectivity of solvent extraction processes with the simplicity and multistage character of a column chromatographic system. While conventional solvent extraction is a liquid-liquid system, extraction chromatographic resins perform the same separations in a solid-liquid system. This potentially allows use of simpler equipment, easier material handling, and reduced capital cost in construction of process equipment.

The resins used in extraction chromatography are prepared by adsorption of a conventional or slightly modified extractant on a macroporous polymeric support. This preparation process immobilizes the extractant while retaining the chemical properties (distribution ratios, selectivities, etc.) of the original liquid solvent system. Thus, extractants known to selectively extract radionuclides in a solvent extraction process [i.e. octyl(phenyl)-N,N-diisobutylcarbamoylmethylphosphine oxide (CMPO) -tributyl phosphate (TBP) mixtures used in the TRUEX process for extraction of TRU elements] can be immobilized on a solid support and will retain their selectivity and extraction capacity (Horwitz et al., 1990).

Chemical Pretreatment of Nuclear Waste for Disposal, Edited by
W.W. Schulz and E.P. Horwitz, Plenum Press, New York, 1995

A number of investigators have prepared extraction chromatographic resins by adsorbing carbamoyl phosphonates or carbamoylmethylene phosphonates onto various substrates and have used these resins for separating actinides and lanthanides from acid solutions. Dihexyl-N,N-diethylcarbamoylmethylene phosphonate (DHDECMP) adsorbed onto Amberlite XAD-4 macroporous resin was used to separate americium, cerium, curium, and californium from fission products (Kimura and Akatsu, 1991a and 1991b) and to recover gram quantities of americium from impurities (Yamada et al., 1982). DHDECMP adsorbed onto Chromasorb 102 beads (Marsh and Simi, 1982) was used to separate lanthanides, uranium, plutonium, and americium from impurities. Kwinta et al. (1985) reported separation of lanthanides from transition elements using DHDECMP adsorbed onto a polychlorotrifluoroethylene packing material. Baker et al. (1980) used DHDECMP adsorbed onto Vydac C8 resin[1] to separate lanthanides from fission products. Dibutyl-N,N-diethylcarbamoylphosphonate (DBDECP) adsorbed onto Microthene[2], a microporous polyethylene (Sarzanini et al., 1988) and on Kel-F 300 powder[3] (Yamamoto et al., 1981) was used to separate plutonium and americium.

Two resins were tested in the present investigation -- a CMPO-impregnated resin prepared in our laboratory and an EIChrom Industries, Inc. (Darien, Illinois) resin (TRU•Spec). These resins extract transuranic elements in (III), (IV), and (VI) oxidation states. The radionuclides loaded onto the resin can be eluted using dilute acid solutions or solutions of complexants.

The extraction chromatographic resins might be used to selectively remove plutonium, americium, uranium, neptunium, and thorium from almost any acidic waste stream generated at the Hanford Site. For example, these materials could be used to treat current acid wastes from the Plutonium Finishing Plant (PFP) such as current acid waste (CAW), acidified carbonate scrub waste (CXP), and low salt wastes. These TRU wastes would be converted to low-level waste and the plutonium recovered from the waste would be recycled. Complexant concentrate waste (CCW), neutralized cladding removal waste (NCRW), and single shell tank waste (SST) are all alkaline wastes that are classified as TRU waste. Sludges precipitated during neutralization of the wastes contain most of the TRU elements originally present in the wastes and can be separated from the salt cake and dissolved in nitric acid. The resulting acid solutions are candidate wastes for treatment with the resins. A simplified flow diagram for these processes is shown in Figure 1.

The primary objective of the present work is to evaluate extraction chromatographic resins for separating actinides from acidic waste streams that are generated at the Hanford Site. Transuranic radionuclides must be removed to levels below the TRU limit (100 nCi/g) for the resins to be considered successful. The treated waste must also meet the requirements for classification as low-level waste. Removal of plutonium, americium, and uranium radioisotopes was studied.

The measurements yielded several types of extraction parameters. Distribution coefficients (Kd's) for actinide tracers were determined under a range of conditions (acid concentration, waste composition, radionuclide concentration, and resin type) expected in the actinide removal system. These batch equilibrium measurements were used for setting experimental parameters for the column tests. Column testing provided loading and elution data that will be used to predict extraction rates for dissolved actinides, capacities, and regeneration efficiencies.

---

[1] Vydac is a trademark (pending registration) of the Separations Group, Inc. Hesperia, California.

[2] Microthene is a registered trademark of the National Distillers and Chemical Corporation, New York, New York.

[3] Kel-F is a registered trademark of the Minnesota Mining and Manufacturing Company, Saint Paul, Minnesota.

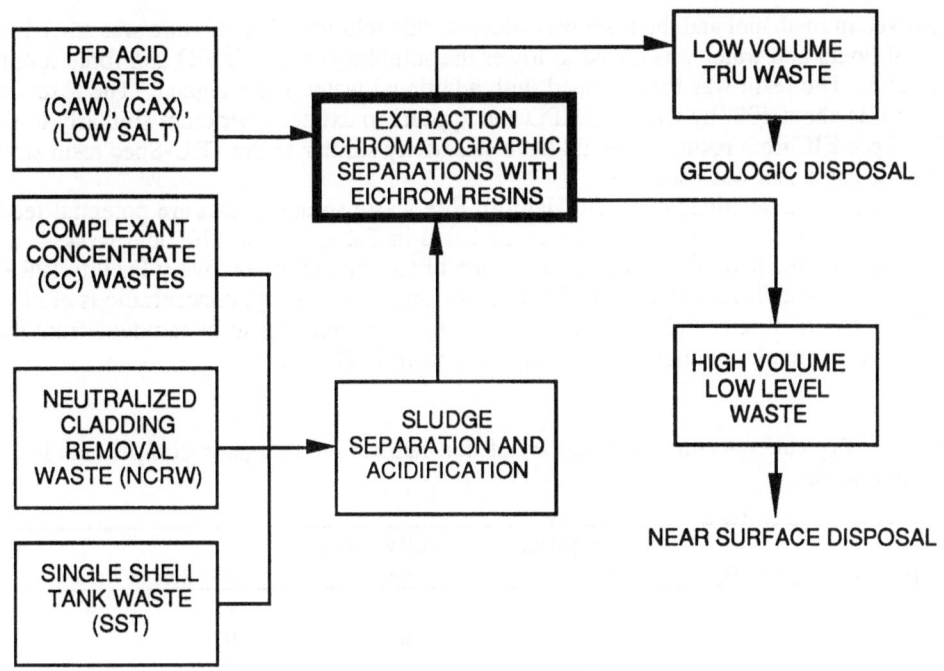

Figure 1. Waste Treatment Flow Diagram.

EXPERIMENTAL

Materials

Extraction chromatographic resins were chosen for their known effectiveness in removing actinides from acidic wastes. Specific removal of actinides is important so that the capacity of the resins is not used up by non-radioactive components of the wastes. It is also important that the actinides have high distribution coefficients for extraction onto the resins since most of the radionuclides will be present in very small concentrations compared to the non-radioactive components. Resins containing the CMPO extractant appear to have the most promising characteristics since they strongly extract actinides in most oxidation states.

A resin, TRU•Spec, manufactured by EIChrom Industries, Inc. (hereafter EIChrom) was tested as well as a laboratory-prepared resin containing CMPO. TRU•Spec resin contains a mixture of CMPO and TBP extractants adsorbed onto a macroporous polymeric support (13, 27, and 60 weight percent, respectively). This resin specifically extracts transuranic elements from acid solutions. Most other metal ions do not interfere with the extraction. Zirconium, bismuth, and some of the rare earth elements are extracted by this resin, but are not expected to be present in significant concentrations in most wastes. Metallic elements that make up the major portion of the wastes (i.e. Na, Al, Fe, Cr, Mg, and Ca) should not be extracted, according to EIChrom literature. The resin can be regenerated by eluting TRU elements with dilute acid or with complexing agents. The useful loading capacity of this resin for americium is about 1 to 2 grams/L of resin bed. The particle size range of resin beads tested is 100 to 125µ.

In addition to the EIChrom resin, a resin prepared in our laboratory was tested. It was prepared by loading CMPO onto macroporous resin beads (Amberlite XAD-16, Rhome & Haas, 20-60 mesh). About 100 g of the resin were washed with water to remove any preservative salts, then air-dried and washed several times with reagent grade methanol. Excess methanol was evaporated by air drying and then the resin was weighed. An amount of crystalline CMPO (M & T Chemicals) equal to 2/3 the weight of the purified resin was dissolved in methanol and the resin was added to this solution. The mixture was stirred for several hours and water was added to lower the solubility of the CMPO and force it onto the resin. The resin was then washed with additional water and weighed. The resulting material is about 40% by weight CMPO and appears to extract americium similarly to the TRU•Spec EIChrom resin. It should have selectivity similar to the TRU•Spec resin since the same extractant was used.

Typical compositions of acidic Hanford Site waste solutions that are potential feeds for extraction chromatography columns are listed in Table 1. The PFP acid waste is the raffinate from the plutonium extraction column of the Plutonium Recovery Facility (PRF). This waste stream is designated the CAW stream and contains high concentrations of nitric acid, aluminum nitrate, and hydrofluoric acid. The radionuclides to be removed from this waste stream are plutonium and americium, as shown in Table 2.

Table 1. Typical concentrations (molar) of non-radioactive inorganic components in the acidified wastes.

Chemical Components	PFP Waste	CC Sludge Waste	NCRW Sludge Waste	SST Sludge Waste
Acid	3.0	2.0	1.5	2.0
Aluminum	0.52	0.10	0.020	0.1
Sodium	0.052	0.30	1.0	0.5
Iron	0.005	0.01	0.010	0.2
Calcium	0.002	--	--	--
Magnesium	0.001	--	--	--
Chromium	--	--	0.010	0.1
Zirconium	--	0.00008	0.15	0.01
Cerium	--	--	--	0.003
Bismuth	--	--	--	0.003
Manganese	--	--	--	0.05
Nitrate	5.0	4.0	--	4.0
Sulfate	--	--	--	0.02
Phosphate	--	--	--	0.02
Fluoride	0.03	--	0.5	0.1
Chloride	0.01	--	--	--

Table 2. Typical radionuclide concentrations (μCi/L) in the acidified wastes.

Radionuclide	PFP Waste	CC Sludge Waste	NCRW Sludge Waste	SST Sludge Waste
Plutonium	1,000	20	1,000	2,000
Americium	5,000	100	100	10,000
Uranium (g/L)	0.2	--	1.0	0.2
Strontium	--	--	1,500	100,000
Cesium	--	20,000	5,000	1,000
Europium	--	500	10	5,000

Typical compositions (Schulz, 1980; Stordeur, 1985) of the three acidified sludge waste solutions are also shown in Table 1. These sludges result from neutralization of PUREX process high-level wastes that have been mixed with various organic complexing agents from subsequent recovery of ^{90}Sr. High concentrations of sodium and aluminum were found in the acidified sludge along with some iron. Cesium, europium, americium and plutonium were the main radioelements present. Strontium was probably also present, but not measured. The acidified NCRW sludge waste contains high concentrations of sodium, zirconium, fluoride, aluminum, iron , and chromium. Cesium, strontium, plutonium, and americium are the principal components. This waste was precipitated during neutralization of solutions from Zircaloy cladding removal. Acidified SST sludge waste is highly variable in composition (Schulz, 1980). Sodium, iron, aluminum, and chromium are the major components of this waste. The major radioelements are strontium, americium, plutonium, europium, cesium, and uranium. On a mole basis, uranium is by far the most abundant radioelement. Synthetic waste solutions containing major components of the waste were prepared for use in the extraction chromatographic experiments using reagent grade chemicals and distilled water. Radioactive tracers (^{238}Pu, ^{241}Am, and ^{232}U) were added to these solutions to measure the effectiveness of their removal by various adsorbents.

Actual samples of CAW waste from the PRF were collected in 1988 for use in confirming the results of testing with synthetic waste solutions. Two samples were treated with columns of the extraction chromatographic resin. The sample compositions are given in Table 3. Both samples were filtered through filter membranes with a pore size of 0.3 μm before pumping them through the columns. Sample 156-2 was also partially neutralized with solid calcium hydroxide to a final acid concentration of 0.5 $\underline{M}$. Apparently, about half of the plutonium precipitated during the neutralization step.

Table 3. Composition of CAW samples used in the tests.

Component	Sample 156-1	Sample 156-2	Neutralized Sample 156-2
Am, μCi/L	3,820	3,980	3,730
Pu, μCi/L (alpha)	58.8	70.8	29.5
Total Alpha, μCi/L	3,880	4,050	3,760
H^+, molar	2.1	2.3	0.5
NO_3^-, molar	4.14	4.43	4.15

Analyses

Radioactive tracer concentrations were measured using liquid scintillation counting (Packard 1500 Tri-Carb liquid scintillation analyzer).

Batch Equilibration Measurements

The batch measurements were performed using 20 mL glass bottles with screw caps to contain the resin-solution mixtures. The bottles were shaken with a variable speed flat bed shaker to speed equilibration. A weighed amount of resin was equilibrated with a measured volume of solution containing known concentrations of tracer and inert components. The acid concentration was adjusted with HNO_3 and measured after equilibration. The time required for equilibration was determined by sampling the solution periodically and measuring the tracer concentration. When a constant tracer concentration was reached, equilibrium had been attained. The concentration of tracer in the solution after equilibration was used to calculate the amount of tracer extracted by the resin. A Kd value (Kd = tracer activity/g resin + tracer activity/mL solution) was then calculated using these data.

Column Extraction Measurements

Chromatography columns (Spectrum Medical Industries, Inc.) were used to contain the resin in the column tests. The resin bed dimensions could be adjusted with these columns. Precision metering pumps (Eldex Laboratories, Inc., model E-120-S) were used to pump solutions through the columns at a known, constant rate. Effluents from the columns were sampled automatically using a fraction collector (Haake Buchler Instruments, Inc., model LC 200) that collected constant volume fractions.

The liquid chromatography columns were filled with resin (added as a water slurry). Solutions containing known concentrations of radioactive tracer and inert components were pumped through the columns at a constant, known flow rate. Samples of the effluents from the columns were counted for tracer concentration. Breakthrough curves were plotted to show tracer concentration in the effluent versus effluent volume. For extraction rate experiments, the flow rate was varied and the effect of flow rate on the tracer concentration in the effluent was evaluated. Resin capacities were determined by measuring the volume of feed solution passed through the column when the tracer is observed in the effluent at 50 % of the feed concentration..

Three column sizes were used. A small column was prepared by adding 2.0 g of water-saturated TRU•Spec resin to a column with an inner diameter of 1.5 cm. This resin bed was 1.9 cm long with a total volume of 3.36 mL and a void volume of 1.44 mL. This column was used to measure loading capacities of the actinides. A second column was prepared by adding 10.0 g of TRU•Spec resin to another 1.5 cm diameter column. The resin bed was 8.6 cm long with a total volume of 15.2 mL and a void volume of 5.58 mL. This column was used to measure extraction kinetics for the actinides. A third column having a bed volume of 53.5 mL (2.5 cm diameter, 10.9 cm long) was used in testing the CAW waste solutions.

The effects of acid concentration, radionuclide concentrations, and concentrations of interfering ions on extraction were determined in both the batch and column tests by changing their values over the specified range of these parameters.

RESULTS AND DISCUSSION

Distribution Coefficients

Identification of Waste Components Affecting Extraction. Statistically-designed screening tests were used to identify important components of the waste solutions that affect actinide extraction on the chromatographic resins. Components (variables) were selected for testing based on measured waste compositions and on expected effects (from EIChrom literature and solvent extraction data). Eight variable components were selected as follows: Fe(III), Cr(III), Zr(IV), Bi(III), Ce(III), and U(VI) nitrates, HF, and H_3PO_4. Since aluminum and sodium nitrates are found in all the wastes, their concentrations were held constant in these tests (0.20 $\underline{M}$ and 0.50 $\underline{M}$, respectively. Nitric acid is known to be an important variable and was therefore not varied in these screening tests, but was held constant at 3.0 $\underline{M}$. The concentrations of the variable components ranged from zero to the highest concentration measured in acidified sludge waste (see Table 1). Twelve solutions were prepared, each having different variable components present. The solutions were prepared according to a 12-run Plackett-Burman screening design (Plackett and Burman, 1946).

The data from these tests were analyzed statistically (Barney and Cowan, 1992) to determine which solution components significantly affect extraction of the actinides. A number of general conclusions can be drawn from the extraction data:

> (1) Extraction of actinides by TRU•Spec resin is significantly decreased by other metal ions that can also be extracted [i.e., Zr(IV), Bi(III), Fe(III) and Ce(III)] because of competition of metal ions for the extractant (CMPO) on the resin.
> (2) Higher nitrate ion concentrations associated with the presence of iron, chromium, and zirconium increase extraction on TRU•Spec resin. Since the actinides extract as uncharged, neutral nitrate salts this behavior was expected.
> (3) When both zirconium and HF are present in the solutions, their effect on actinide extraction is significantly lowered due to formation of very stable ZrF_x

complexes. In actual wastes, fluoride will always be present with zirconium, so that the effect of both on actinide extraction will likely be small.

Effects of Acid, Aluminum Nitrate, and Ferric Nitrate on Americium Extraction. The major components of the acidified wastes are H^+, Na^+, Al^{3+}, Fe^{3+}, NO_3^-, and F^- according to Table 1. Acid and nitrate are known to strongly affect americium extraction by CMPO (Schulz and Horwitz, 1988). Nitric acid and metal ions (M^{3+}) compete with americium for the CMPO extractant according to the following equations:

$$Am(NO_3)_3 \cdot (CMPO)_3(HNO_3)_m + (3n-m)HNO_3 = Am^{3+} + 3NO_3^- +$$

$$3(CMPO) \cdot (HNO_3)_n$$

where n = 0 to 2 and m = 0 to 3, and

$$Am(NO_3)_3 \cdot (CMPO)_3(HNO_3)_m + M^{3+} = Am^{3+} +$$

$$M(NO_3)_3 \cdot (CMPO)_3(HNO_3)_m.$$

Sodium and aluminum ions are not extracted but they are present as nitrate salts and will increase the concentration of nitrate in solution. Iron was observed to lower americium extraction in the screening tests reported above. Fluoride does not appear to be an important factor since it will be complexed strongly in solution by Al^{3+}, Fe^{3+}, and other metal ions. Because of these considerations detailed measurements of the effects of H^+, $Al(NO_3)_3$, and $Fe(NO_3)_3$ concentrations on americium extraction was completed.

Since americium is extracted to a lesser extent than other actinides in the waste and generally has the greatest activity of actinides in the waste, it was assumed that conditions for successful extraction of americium would be even more favorable for the other actinides. This assumption will be tested in future work.

Batch equilibrations of solutions containing a range of HNO_3, $Al(NO_3)_3$, and $Fe(NO_3)_3$ concentrations with the TRU•Spec resin were performed. Nitric acid concentration ranged from 0.1 to 4.0 $\underline{M}$, $Al(NO_3)_3$ concentrations ranged from 0 to 0.5 $\underline{M}$, and $Fe(NO_3)_3$ concentration ranged from 0 to 0.2 $\underline{M}$. In addition, each solution contained 0.5 $\underline{M}$ $NaNO_3$. One gram of resin (pre-equilibrated with 0.1 $\underline{M}$ HNO_3 - 0.5 $\underline{M}$ $NaNO_3$ solution for two hours) was contacted with 20 mLs of solution containing 0.017 $\mu Ci/mL$ of ^{241}Am. The mixtures were equilibrated for 24 hours, filtered through 1.0 μm pore size Millipore Teflon membrane filters, and americium concentrations in the filtrate were determined by liquid scintillation counting.

Distribution coefficients for americium extraction [Kd(Am) values] were calculated and are plotted in Figures 2, 3, and 4 versus nitric acid concentration. Figure 2 shows Kd(Am) values for different $Al(NO_3)_3$ concentrations when no iron is present in solution. As expected from the extraction equilibrium equations, nitrate [from $Al(NO_3)_3$] strongly increases americium extraction and acid decreases americium extraction. The effect of nitrate is much greater at low acid concentrations. Figures 3 and 4 show a large decrease in Kd(Am) when iron is present. The effect of iron is greater at higher acid concentration. Aluminum nitrate concentration has almost no effect when iron is present. This may be due to the fact that both americium extraction and iron extraction are enhanced by increasing nitrate concentration. Since iron competes with americium extraction these two effects cancel each other.

The distribution coefficient data indicate that americium extraction from the waste solutions by TRU•Spec resin would be most efficient at low acid concentration. Based on expected iron concentrations in the wastes, Kd(Am) values ranging from about 35 to >100 mL/g can be achieved if the acid concentration is maintained in the range of 0.1 to 0.5 $\underline{M}$. The other actinides will be more strongly extracted than the americium. These values are probably high enough to design a practical process for separating actinides from the acidified wastes.

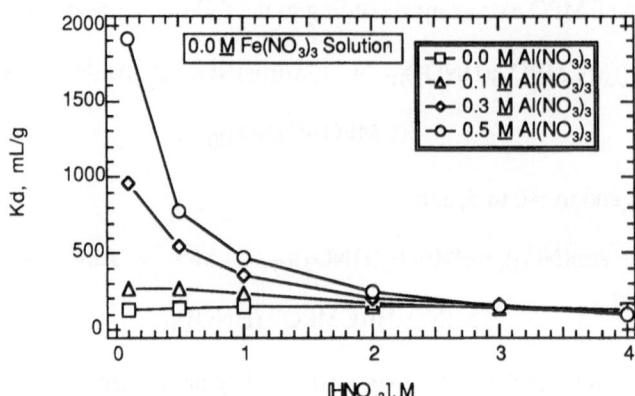

Figure 2. Effects of aluminum nitrate and nitric acid concentrations on americium extraction by TRU•Spec extraction chromatographic resin with no iron present.

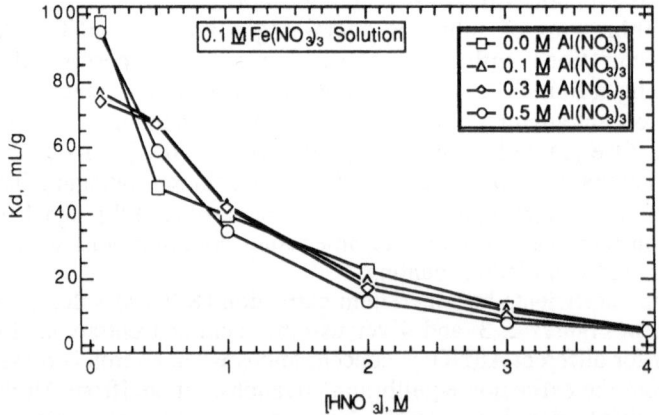

Figure 3. Effects of aluminum nitrate and nitric acid concentrations on americium extraction by TRU•Spec extraction chromatographic resin at 0.1 $\underline{M}$ Fe(NO3)3.

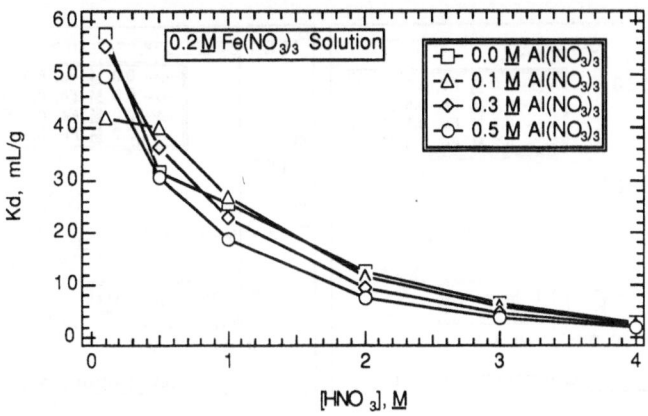

Figure 4. Effects of aluminum nitrate and nitric acid concentrations on americium extraction by TRU•Spec extraction chromatographic resin at 0.2 $\underline{M}$ Fe(NO$_3$)$_3$.

Effects of Uranium and Bismuth on Americium Extraction. The quantities of uranium and bismuth in the wastes will be small compared to the major components examined in the previous section. They can still have a significant effect on americium extraction with TRU•Spec resin because they are extracted strongly by CMPO. Both UO_2^{2+} and Bi^{3+} ions are expected to compete with americium for the CMPO extractant and thus lower extraction distribution coefficients. As was the case for previous Kd measurements, americium extraction was used as an indicator for extraction of the other significant actinides since it is extracted more weakly.

Separate batch equilibrations of solutions containing a range of uranyl nitrate and bismuth nitrate concentrations with the TRU•Spec resin were completed. These solutions also contained several different concentrations of HNO$_3$, Al(NO$_3$)$_3$, and Fe(NO$_3$)$_3$ so that their effect on americium extraction could be determined. Concentration ranges for UO$_2$(NO$_3$)$_2$ and Bi(NO$_3$)$_3$ were 0.01 to 2.0 g/L and 0.002 to 0.05 $\underline{M}$, respectively. One gram of resin (pre-equilibrated with 0.1 $\underline{M}$ HNO$_3$ - 0.5 $\underline{M}$ NaNO$_3$ solution for two hours) was contacted with 20 mLs of solution containing 0.017 µCi/mL of [241]Am. The mixtures were equilibrated for 24 hours, filtered through 1.0 µm pore size Millipore Teflon membrane filters and americium concentrations in the filtrate were determined by liquid scintillation counting.

Americium distribution coefficients are plotted in Figures 5 and 6 for ranges of HNO$_3$, UO$_2$(NO$_3$)$_3$, Bi(NO$_3$)$_3$, Fe(NO$_3$)$_3$, and Al(NO$_3$)$_3$ concentrations. Figure 5 shows that uranium concentrations greater than about 0.1 g/L can significantly reduce americium extraction. The effect of uranium is not as pronounced when iron is present. As shown previously, Fe^{3+} and H^+ lower americium extraction and NO_3^- increases it. Bismuth seriously interferes with americium extraction as indicated in Figure 6. Even at a bismuth concentration as low as 0.002 $\underline{M}$, americium Kd values are about a factor of two lower than values obtained under the same conditions except with no bismuth present. For those wastes containing bismuth (from the Bismuth Phosphate separation process used at the Hanford Site) interference of bismuth in actinide extraction may have to be addressed by either complexing it or removing it from the dissolved wastes before actinide removal.

Comparison of TRU•Spec and Laboratory-Prepared Resins. The extraction of americium with TRU•Spec resin was compared with laboratory-prepared resin containing about 40 percent by weight (dry basis) CMPO adsorbed onto Amberlite XAD-

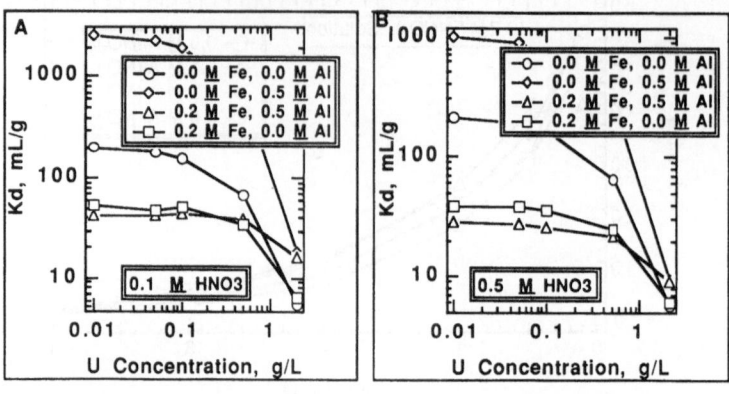

Figure 5. Effects of uranium on americium extraction by TRU•Spec extraction chromatographic resin at two different acid, iron and aluminum concentrations.

16 (see Experimental Section for preparation). The same procedure was used to measure Kd(Am) values for the laboratory-prepared CMPO-resin. Portions of the same 72 americium-spiked solutions were used to equilibrate the resin where HNO_3, $Al(NO_3)_3$, and $Fe(NO_3)_3$ concentrations were varied. Americium Kd values for the resins are compared in Figure 7. Values for the TRU•Spec resin (filled squares) and CMPO-resin (open circles) are shown for the 72 experiments and the values for each experimental solution composition are connected by vertical lines. In most experiments no line is apparent because the two values are virtually identical. For the low-numbered experiments (1 to 12), where the acid concentration is 0.1 $\underline{M}$, TRU•Spec resin gives generally higher Kd(Am) values. For the rest of the experiments, however, the Kd(Am) values are almost the same for the two resins, within experimental error.

Based on these results, it appears that a resin with extraction properties very similar to

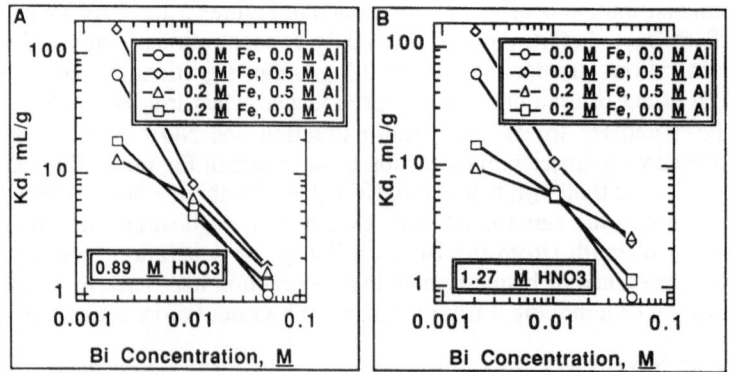

Figure 6. Effects of bismuth on americium extraction by TRU•Spec extraction chromatographic resin at two different acid, iron and aluminum concentrations.

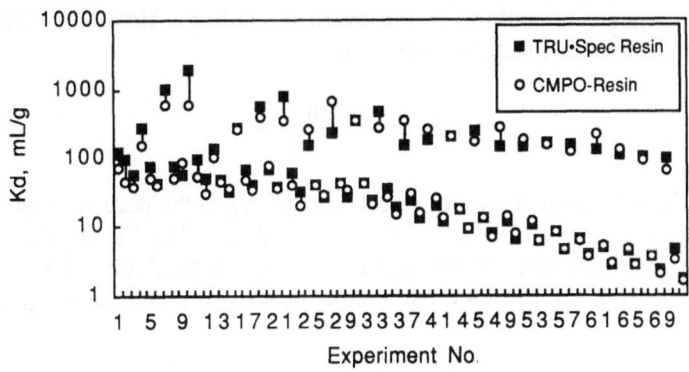

Figure 7. Comparison of Kd(Am) values for TRU•Spec resin and CMPO-Resin in 72 Solutions for different compositions. Each pair of data points was measured in identical solutions.

TRU•Spec can be easily prepared in the laboratory. The differences in the composition of TRU•Spec and CMPO-resin are: (1) 13 and 40 percent CMPO by weight, respectively, (2) 27 and 0 percent TBP, respectively, and (3) acrylic ester polymer support and macroporous polystyrene-divinyl benzene copolymer support, respectively. Despite these significant differences, the resins extract americium very similarly.

Column Tests

Americium Capacity Test. The small column containing 2.0 g of TRU•Spec resin as described in the Experimental section was used to measure americium extraction capacity. This small column was used so that americium breakthough curves could be observed in a reasonable length of time. The disadvantages of using a small column are (1) increased probability of channeling and (2) relatively large volumes of "dead pores" or stagnant solution in the column. These conditions lead to premature breakthrough and increased tailing in the breakthrough curves. These effects will not significantly affect calculations of column capacity, however.

Two ^{241}Am-spiked solutions with different acid concentrations were passed through the column to measure americium loading capacity of the resin and the shape of americium breakthrough curves. The first solution had the following composition: 0.10 $\underline{M}$ HNO$_3$, 0.50 $\underline{M}$ NaNO$_3$, 0.20 $\underline{M}$ Al(NO$_3$)$_3$, and 0.20 $\underline{M}$ Fe(NO$_3$)$_3$. The second solution had the same composition except 0.50 $\underline{M}$ HNO$_3$. Both solutions were spiked with 0.17 µCi/mL ^{241}Am. Prior to running the spiked solutions through the column, it was pre-equilibrated with unspiked solution of the same composition. The solutions were passed upflow through the column at a flow rate of 0.5 mL/min (column residence time was about 2.9 min) and the effluent was collected in 5 milliliter fractions in plastic test tubes using the automatic fraction collector. One milliliter samples were withdrawn from each tube and analyzed for americium using liquid scintillation counting. The americium and iron extracted from the first solution were eluted using 0.5 $\underline{M}$ 1-hydroxyethyl-1,1-diphosphonic acid (HEDPA) and the column was then pre-equilibrated with unspiked solution having the composition of the second solution before running the second spiked solution.

Both breakthrough curves are shown in Figure 8 where C/Co is the ratio of americium concentration in the effluent (C) and feed (Co). Because of the small column size, the curves show premature breakthrough and extensive tailing. However, Kd(Am) values calculated from these curves using the equation,

$$Kd(Am) = \frac{V_{50} - V_p}{M},$$

(where V_{50} is the volume at 50 percent breakthrough, V_p is the pore volume, and M is the mass of resin) were close to the measured batch values. The batch values for 0.10 $\underline{M}$ HNO$_3$ and 0.5 $\underline{M}$ HNO$_3$ were 50 mL/g and 38 mL/g, respectively. The corresponding values calculated from the breakthough curves are 57 mL/g and 32 mL/g, respectively.

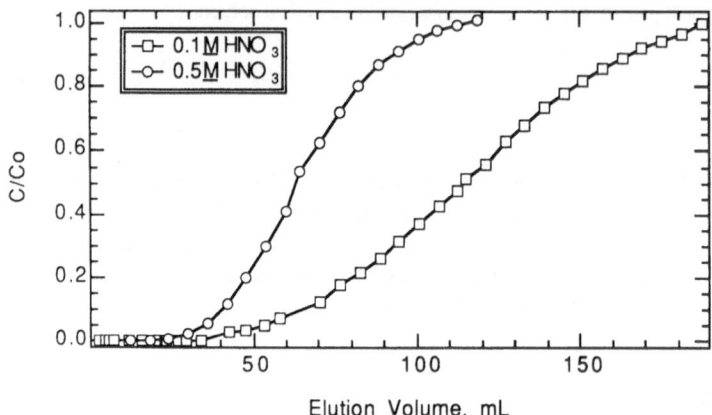

Figure 8. Effect of nitric acid concentration on americium breakthrough curves for solutions containing 0.50 $\underline{M}$ NaNO$_3$, 0.2 $\underline{M}$ Al(NO$_3$)$_3$, and 0.20 $\underline{M}$ Fe(NO$_3$)$_3$.

These results show that americium extraction reactions are relatively rapid and are at near equilibrium in the column. The resin capacities for americium extraction (at 50 percent breakthrough) in terms of liters of solution processed per liter of resin are about 34 and 19 for 0.10 $\underline{M}$ HNO$_3$ and 0.50 $\underline{M}$ HNO$_3$, respectively. These values would, of course, be much higher at Fe(NO$_3$)$_3$ concentrations lower than 0.20 $\underline{M}$ since the Kd(Am) values are higher.

Iron is obviously extracted in these experiments since the color of the resin changes from white to a light orange-yellow color after contact with the simulated waste solutions. Iron saturation of the extractant is reached very quickly because the concentration of iron in the solution is high and the Kd(Fe) is relatively low.

Extraction and Elution Kinetics. The rates of actinide extraction and elution reactions in the resin bed are important factors in determining the maximum flow rates through the bed and the minimum volume of the bed for practical application to removal of actinides from the waste solutions. Rate measurements of actinide loading were attempted by passing solutions spiked with [241]Am and [238]Pu through a column of TRU•Spec resin at variable flow rates. The effluent concentrations of tracer after reaching a steady-state value were compared for different flow rates. Low flow rates result in longer column residence times for the solutions and should yield lower steady-state concentrations of tracer.

After each loading test, the actinide was eluted using a complexant solution and the tracer was measured in the eluate. If elution reactions are slow, low flow rates should lead to higher initial effluent tracer concentrations and more tracer should be removed from the resin for a given volume of eluant. Steady-state effluent concentrations of tracer will not be achieved, but should continue to decrease with increasing effluent volume.

The large column described in the Experimental Section containing 10.0 g of TRU•Spec resin was used in these tests. This column largely eliminated channeling effects observed with the smaller column.

Americium loading tests were performed using two different solution compositions. The first solution contained 0.1 $\underline{M}$ HNO_3, 0.5 $\underline{M}$ $NaNO_3$, 0.2 $\underline{M}$ $Al(NO_3)_3$, and 0.2 $\underline{M}$ $Fe(NO_3)_3$. This solution was spiked with 0.089 µCi/mL of [241]Am. The second solution had the same composition as the first except that the HNO_3 concentration was 0.5 $\underline{M}$ and the americium spike concentration was 0.083 µCi/mL. Six different flow rates were used to feed the synthetic waste solutions to the bottom of the columns -- 0.5, 1.0, 2.0, 3.0, 4.0, and 5.0 mL/ min. These flow rates cover the range of (flow rate)/(resin volume) normally used for large-scale ion exchange applications. These same flow rates were used for americium elution using 0.5 M HEDPA.

Americium breakthrough curves for loading and elution are plotted in Figure 9. The loading curves show an initial peak concentration in the effluent at about 15 mL that is due to mixing of the feed solution with the HEDPA eluant solution left in the column from the previous elution step. A steady-state concentration in the effluent is not reached until most of the HEDPA is flushed out of the column. Several of the loading curves did not reach a constant americium concentration before 50 mLs had passed through the column (about 8.6 void volumes). There is no clear trend in the effect of flow rate on effluent americium concentration. Other effects appear to be more important, such as completeness of eluting americium from the resin. Low initial americium concentrations in the effluent (an indication of effective removal of americium from the resin during the previous elution) lead to low concentrations near the end of the loading cycle.

Decontamination factors (DF's) for the 0.1 $\underline{M}$ HNO_3 solution are somewhat higher (more complete removal) than for the 0.5 $\underline{M}$ HNO_3 solution. They range from about 4,800 to 42,000 and 3,000 to 12,000, respectively. The lower DF values for the 0.5 $\underline{M}$ HNO_3 solution are due to lower Kd(Am) values at increased acidity (see Figure 4). Elution of the americium is nearly complete at all the flow rates, but as might be expected, is slightly more complete at low flow rates.

The two solution compositions used in the americium loading tests were also used in plutonium loading tests. The solutions containing 0.1 $\underline{M}$ HNO_3 and 0.5 $\underline{M}$ HNO_3 were both spiked with [238]Pu at an activity level of 0.017 µCi/mL. Again, six different flow rates were used to feed the synthetic waste solutions to the bottom of the column and to elute the column-- 0.5, 1.0, 2.0, 3.0, 4.0, and 5.0 mL/ min. The eluant used for plutonium was a 0.5 $\underline{M}$ solution of ammonium citrate adjusted to a pH of 5.

Loading and elution breakthrough curves for plutonium are shown in Figure 10. As for americium, the initial increase in the effluent plutonium concentration was due to eluant solution left in the column from the previous elution step. Steady-state plutonium concentrations were reached quickly in most cases. There is also a trend to higher concentrations at higher flow rates. This is expected if the extraction rate is slow compared with the solution residence time in the column.

Plutonium DF's were generally higher for the 0.5 $\underline{M}$ HNO_3 solution than for the 0.1 $\underline{M}$ HNO_3 solution. Values for DF ranged from about 700 to 67,000 and 660 to 4,300 for the two solutions, respectively. Ammonium citrate appears to be an efficient, rapid eluant for the plutonium. Flow rate had very little effect on the elution curves.

Uranium loading [as U(VI)] was examined with just one waste solution composition. The solution was spiked to an activity level of 0.02 µCi/mL of [232]U. Five flow rates were used to feed the solution to the bottom of the column -- 0.5, 1.0, 3.0, 4.0, and 5.0 mL/min. These rates were also used for uranium elution with 0.5 $\underline{M}$ ammonium citrate adjusted to pH 5. The loading and elution breakthrough curves are shown in Figure 11. A steady-state effluent concentration of uranium was reached after about 30 mLs had eluted. There is no clear trend in the effect of flow rate on uranium effluent concentration. Decontamination factors ranged from about 1,200 to 2,600.

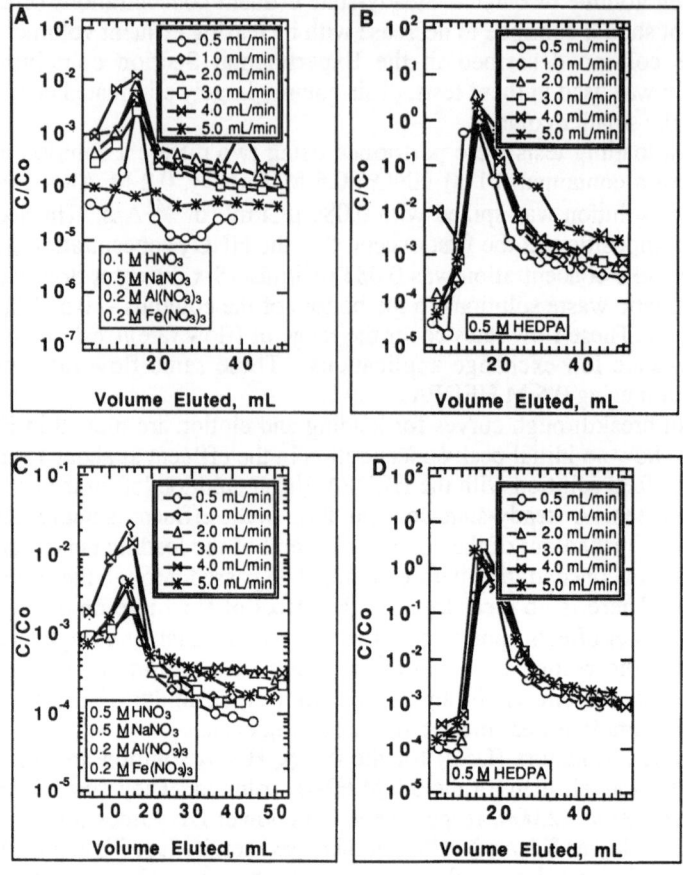

Figure 9. Effect of flow rate on americium breakthrough curves for loading (A and C) and elution (B and D) of two different synthetic waste compositions.

Extraction of Plutonium and Americium from Actual Waste. Removal of americium and plutonium from an acidic waste stream generated at the Plutonium Finishing Plant on the Hanford Site was tested using the TRU•Spec extraction chromatographic resin. The waste stream tested was the CAW stream from the Plutonium Recovery Facility (PRF) solvent extraction process. The objectives of the tests were to determine the ability of the resin to remove americium and plutonium from the CAW waste and to compare these test results with those from a TRUEX Process Pilot Plant run completed earlier in this laboratory using the same waste (Naser et al., 1988).

Actual samples of CAW waste were collected in 1988 during a PRF run and have been stored in the laboratory since then. Five four-L samples were collected, but only two of these, Samples 156-1 and 156-2, were used in the tests. Sample 156-1 was tested without any pre-treatment, except filtering it through a filter membrane with a pore size of 0.3 μm. Sample 156-2 (about 3 L of solution) was filtered and the acid was partially neutralized with solid calcium hydroxide (200 g) to a final acid concentration of 0.5 $\underline{M}$.

The results of the two test runs are given in Figure 12. Values of C/Co for plutonium and americium are plotted versus cumulative effluent volume. The results of the tests show that americium and plutonium can be removed from actual CAW waste to low activities. The column capacity is limited by americium loading since plutonium is extracted to a

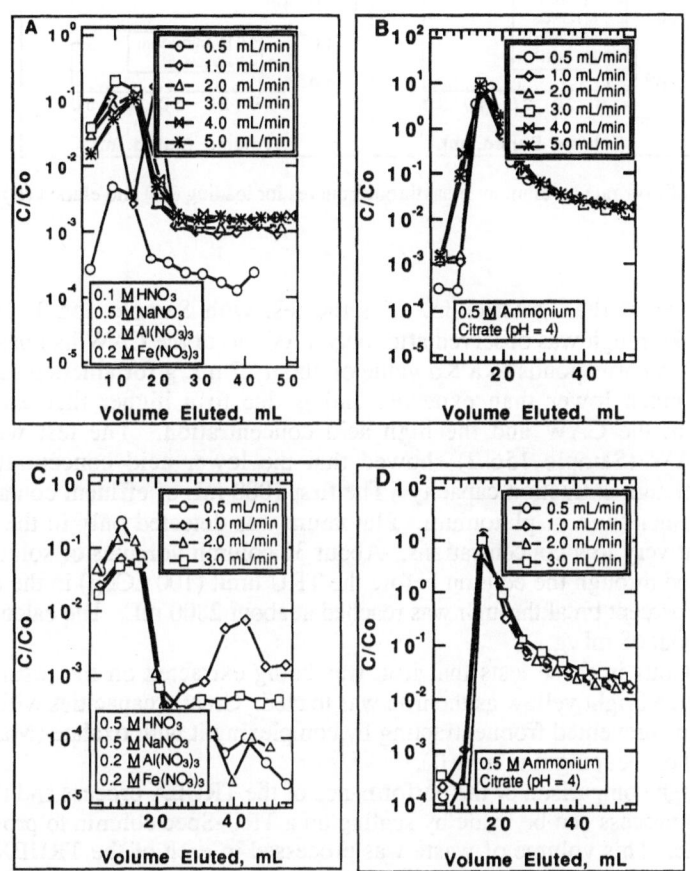

Figure 10. Effect of flow rate on plutonium breakthrough curves for loading (A and C) and elution (B and D) of two different synthetic waste compositions.

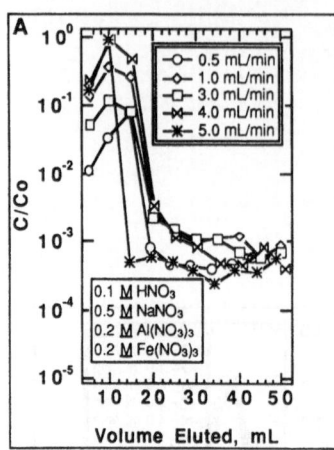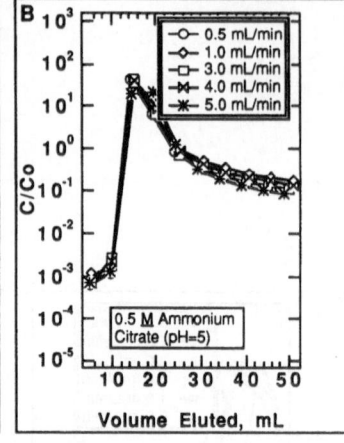

Figure 11. Effect of flow rate on uranium breakthrough curves for loading (A) and elution (B) of a synthetic waste solution.

much greater extent than americium. For the test with Sample 156-1, fifty percent americium breakthrough was observed after only 1100 mL of the CAW had passed through the column. This corresponds to a Kd value of about 27 mL/g for americium adsorption. This value is much lower than expected and is due to a higher than expected iron concentration in the CAW and the high acid concentration. The test with partially neutralized CAW (Sample 156-2) showed that the lower acid concentration greatly improved americium extraction capacity. The first 1000 mL of effluent contained almost no detectable americium or plutonium. Plutonium was detected only in the last several fractions and at very low concentrations. About 32 column volumes of solution (~ 1700 mL) were passed through the column before the TRU limit (100 nCi/g) in the effluent was reached. Fifty percent breakthrough was reached at about 2800 mL. The calculated Kd for this test was about 68 mL/g.

It was obvious in these tests that iron was being extracted on the resin. The resin gradually turned a bright yellow as the iron was loaded. Greater capacities will be obtained if the iron can be prevented from extracting by complexing it with oxalate (Muscatello and Navratil, 1986) or reducing it to iron(II).

A very rough comparison of the performance of the TRUEX process and the EIChrom resin treatment process can be made by scaling up a TRU•Spec column to process 10 L of the CAW waste. This volume of waste was processed in each of the TRUEX Pilot Plant runs. Comparison is made between Run 4 of the Pilot Plant tests and the scaled up EIChrom column (250 milliliters with neutralized feed) in Table 4. The removal of plutonium and americium appear to be comparable for both processes. The volume of extractant and processing space required was significantly less for the EIChrom process. Secondary waste volumes are also less with the EIChrom process.

TRU•Spec Stability Test. The chemical and radiolytic stability of resin used to remove actinides from the wastes is an important consideration in assessing the practicality of its use. Continuous loading and elution cycles with the extraction chromatographic resin bed could lead to hydrolysis and radiolysis of the extractant and/or the polymer support. Also, the adsorbed extractant might be slowly leached from the surface of the support by dissolution or physical displacement into the solutions passing through the resin bed. These processes should degrade the performance of the resin over time.

Since the 10.0 g TRU•Spec resin column was used repeatedly in these tests (over 36 loading and elution cycles), a comparison of the column's performance after this use with the performance of the freshly-prepared column was attempted. Americium loading and elution curves using the same feed solution [0.1 $\underline{M}$ HNO$_3$, 0.5 $\underline{M}$ NaNO$_3$, 0.2 $\underline{M}$ Al(NO$_3$)$_3$, and 0.2 $\underline{M}$ Fe(NO$_3$)$_3$] were measured. The eluant was 0.5 $\underline{M}$ ammonium citrate (pH = 5) rather than the HEDPA used in the earlier tests with fresh resin.

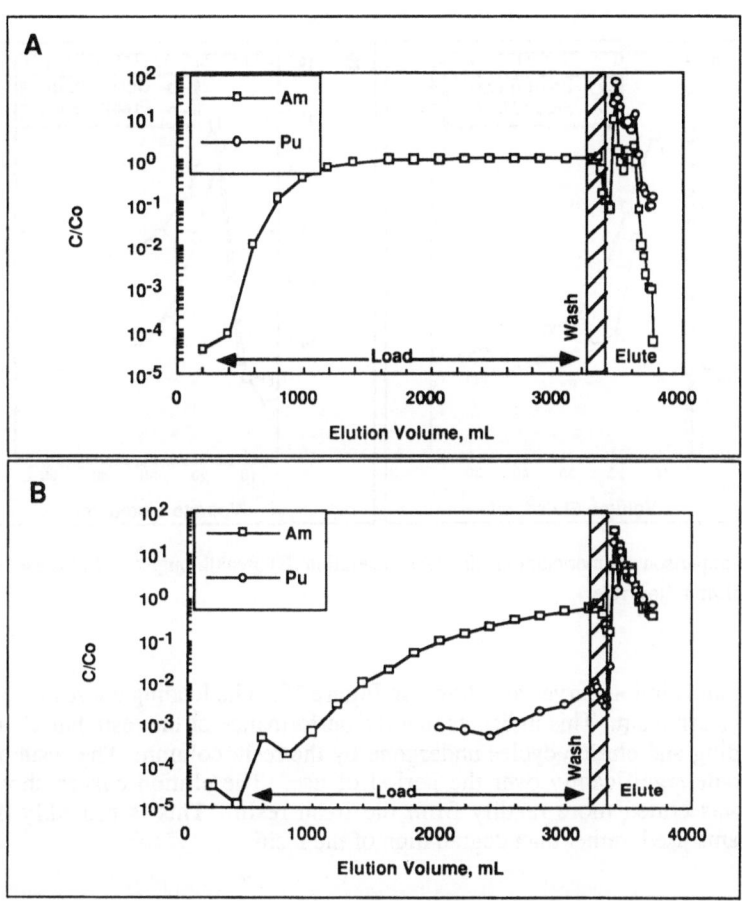

Figure 12. Americium and plutonium loading and elution curves for CAW samples 156-1 (A) and 156-2 neutralized (B).

Table 4. Comparison of feed and effluent activities and volumes.

Process Step	Process Parameter	TRUEX	EIChrom
Extraction	Feed Am Activity, μCi/L	6,490	3,730
	Feed Pu Activity, μCi/L	1,820	29.5
	Effluent Am Activity, μCi/L	51.4	71.4
	Effluent Pu Activity, μCi/L	0.82	<0.02
	Wash Volume/Feed Volume	0.21	0.05
	Volume of Extractant Required, L	4.8	0.25
Stripping	Strip Solution Volume/Feed Volume	0.7	0.2

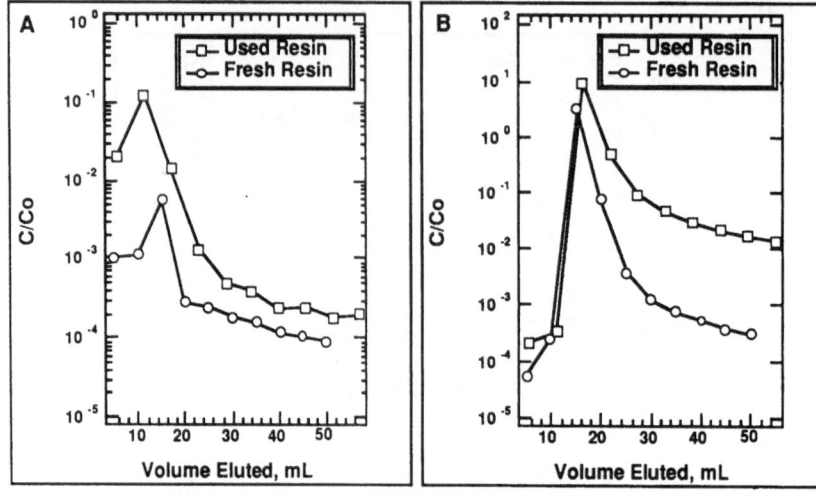

Figure 13. Comparison of americium loading (A) and elution (B) breakthrough curves for used and fresh resin at 1.0 mL/min flow rate.

Loading and elution curves are shown in Figure 13. The loading curves are similar for the used and fresh resin. This indicates that the performance of the resin has changed little over the loading and elution cycles undergone by the resin column. The resin apparently did not degrade significantly over the period of use. The elution curves show that the americium was eluted more readily from the fresh resin. This is probably due to the different eluants used, rather than degradation of the resin.

APPLICATION TO LARGE-SCALE PROCESSING

Extraction chromatography is a possible alternative to the TRUEX liquid/liquid solvent extraction process proposed for acid wastes at the Hanford Site. The proposed TRUEX solvent extraction process uses CMPO/TBP in a normal paraffin hydrocarbon diluent as the organic phase in centrifugal contractors. Concerns about the use of high speed rotating equipment in an acidic, radioactive environment and the volumes of secondary waste generated have prompted investigation of alternatives for removal of TRU elements from acidified wastes.

The present concept for the use of resin columns is to have one column loading and a second column being stripped. To be an attractive alternative, the columns should be of reasonable size to fit within the space of a centrifugal contractor system and generate less stripping solution than does the liquid/liquid extraction process. The present TRUEX process flow sheets require a strip flow of 1/8 of the feed flow.

As the chromatography resins appear to require about 5 column volumes to wash and strip the TRU elements, a column loading of at least 40 volumes is desirable to compete with solvent extraction. The concentration of TRU elements in the tank wastes is very low. Only a small fraction of the resin capacity will be loaded with TRU elements when the equilibrium distribution for the resin is reached. This distribution ratio as, measured in the presence of the competing ions such as iron, bismuth, and uranium, determines the maximum column loading. Although lower values may be acceptable, a minimum Kd(Am) of 50 (Am is the limiting TRU element for most Hanford Site wastes) would make extraction chromatography columns very attractive from a secondary waste standpoint.

The column size will be directly proportional to the loading flow rate. If 200 mm columns X 2 m long are used, a 20 L/min processing rate is equivalent to 1/3 column volume per minute. This is the highest flow rate being tested with these resins and would result in a very compact process which should use less cell space than an equivalent solvent extraction system. This flow rate and column diameter require a 20-50 mesh resin bead size to prevent excessive pressure drop in the column. Higher rate processes could use multiple columns and still retain criticality control using a geometrically favorable design.

A wide range of tank waste compositions exist at Hanford and can not be processed for TRU element removal with one set of operating conditions. The TRU element extraction process must have flexibility to accommodate changing requirements and waste types. Extensive testing remains to demonstrate extraction chromatography with complete range of tank waste proposed for TRU element removal. An extraction chromatography process for ^{90}Sr removal is also needed. Strontium-90 removal is also required for most of the tank wastes being considered for TRU removal. If solvent extraction is required for ^{90}Sr removal, it is likely it will also be used for TRU element removal.

CONCLUSIONS

Based on testing completed thus far, the EIChrom TRU•Spec resin and the laboratory-prepared CMPO-resin can be used to remove actinides from Hanford Site acidic waste streams to levels that are considered non-TRU wastes. Both plutonium and americium decontamination factors are high enough to convert the current acid wastes and acidified sludge wastes to non-TRU waste. Uranium is also extracted, but has a low enough activity that it will not affect the waste classification. Actinide extraction is decreased by the presence of metal ions that compete for the CMPO complexant in the resin. Dissolved iron is an interfering ion that is present in high concentrations in several wastes. Bismuth and uranium may also interfere if present in high enough concentrations. Nitrate ion increases extraction of the actinides due to nitrate complex formation, while acid decreases extraction because of competition for the CMPO extractant. Acid concentration should be in the range of about 0.1 to 0.5 $\underline{M}$ for the extraction of americium to be effective.

Extraction of actinides is rapid by the TRU•Spec resin as shown by column flow-through tests where the flow rate was varied. Concentrations of actinides in the column effluents were not greatly changed by the column feed rate. Elution with ammonium citrate is also rapid and nearly complete after about four pore volumes of the elutant passes through the column. The capacity of the TRU•Spec resin for extracting americium, as measured by column breakthrough measurements, are near those predicted from the equilibrium Kd values. The americium capacities (at 50 % breakthrough) in terms of liters of solution processed per liter of resin was about 34 and 19 for 0.10 $\underline{M}$ HNO$_3$ and 0.50 $\underline{M}$ HNO$_3$ solutions, respectively. These capacities are for solutions containing 0.2 $\underline{M}$ Fe^{3+} which lowers the capacity significantly.

A bench scale test where actual CAW waste was passed through columns of the TRU•Spec resin showed that plutonium and americium can be effectively removed from the waste to very low levels. A 250 mL column of the TRU•Spec resin can treat 10 L of waste (neutralized to 0.5 $\underline{M}$ H$^+$) with results comparable to treatment with a TRUEX process pilot plant solvent extraction system. Even better removal of americium would likely have been achieved if the iron had been complexed with oxalate to prevent it from extracting and competing with americium for the CMPO extractant.

A laboratory-prepared resin with CMPO as the extractant adsorbed onto Amberlite XAD-16 had americium extraction properties very similar to the TRU•Spec resin.

The TRU•Spec resin appears to be stable over at least 36 loading and elution cycles. Americium extraction and elution curves did not significantly change after extended use.

REFERENCES

Baker, J. D., R. J. Gehrke, R. C. Greenwood, and D. H. Meikrantz, 1981, Rapid separation of individual rare-earth elements from fission products, *Radiochimica Acta*. 28:51.

Barney, G. S. and R. G. Cowan, 1992, "Separation of Actinide Ions from Radioactive Waste Solutions Using Extraction Chromatography," WHC-SA-1520-FP, Westinghouse Hanford Company, Richland, Washington.

Horwitz, E. P., M. L. Dietz, D. M. Nelson, J. J. LaRosa, and W. D. Fairman, 1990, Concentration and separation of actinides from urine using a supported bifunctional organophosphorus extractant, *Analytica Chimica Acta*. 238:263-271.

Kimura, T. and J. Akatsu, 1991a, Extraction chromatography in the DHDECMP-HNO$_3$ system. II. Extraction behavior of Ce(III) and Am(III) with the DHDECMP/XAD-4 resin, *J. Radioanal. and Nucl. Chem.* 149:13.

Kimura, T. and J. Akatsu, 1991b, Extraction chromatography in the DHDECMP-HNO$_3$ system. I. Characteristics of the DHDECMP/XAD-4 resin on separation of trivalent actinide elements, *J. Radioanal. and Nucl. Chem.* 149:25.

Kwinta, J., P. David, and G. Metzger, 1985, Preliminary separation of the lanthanide group using DHDECMP, in "Actinide/Lanthanide Separations," G. R. Choppin, J. D. Navratil, and W. W. Schulz (Eds.), World Science Publishing Co., Philadelphia, PA.

Marsh, S. F. and O. R. Simi, 1982, Applications of DHDECMP extraction chromatography to nuclear analytical chemistry, in "Analytical Chemistry in Nuclear Technology," M. S. Lyon (Ed.), Ann Arbor Science Publishers, Ann Arbor, MI.

Muscatello, A. C., and J. D. Navratil, 1986, Americium removal from nitric acid waste streams, in "International Conference on Separations Science and Technology," Litarvan Literature.

Naser, A. J., G. S. Barney, G. A. Escobar, and M. J. Duchsherer, 1988, "TRUEX Process Demonstration Tests with Plutonium Finishing Plant Wastes, " SD-WM-DTR-020, Westinghouse Hanford Company, Richland, Washington.

Plackett, R. L. and J. P. Burman, 1946, The design of optimum multifactorial experiments, *Technometrics.* 3:305-325.

Sarzanini, C., E. Mentasti, P. Benzi, P. Volpe, P. Spezzano, and R. Giacomelli, 1988, Liquid-liquid extraction and selective sorption of actinides by use of carbamoyl-phosphonic ligands, *Radiochimica Acta*, 43:153.

Schulz, W. W., 1980, "Removal of Radionuclides from Hanford Defense Waste Solutions," RHO-SA-51, Rockwell Hanford Operations, Richland, Washington.

Schulz, W. W. and E. P. Horwitz, 1988, The TRUEX process and the management of liquid TRU waste, *Separation Science and Technology.* 23:1191.

Stordeur, R. T., 1985, "Hanford Defense Waste Disposal Alternatives: Engineering Support Data for the Hanford Defense Waste - Environmental Impact Statement," RHO-RE-ST-30P, Rockwell Hanford Operations, Richland, Washington,.

Yamada, W. I., L.L. Martella, and J. D. Navratil, 1982, Americium recovery and purification using a combined anion exchange-extraction chromatography process, *J. Less-Common Metals*. 86:211.

Yamamoto, M., K. Komura, and M. Sakanoue, 1981, A simple sequential separation method of Pu and Am by anion exchange and extraction chromatography," *Radiochimica Acta*. 29:205.

CHEMICAL MECHANISMS FOR GAS GENERATION IN TANK 241-SY-101

D.M. Strachan[1], L.R. Pederson[1], S.A. Bryan[1],
E.C. Ashby[2], C.L. Liotta[2], E.K. Barefield[2],
H.M. Neumann[2], F. Doctorovitch[2], A. Konda[2], K. Zhang[2],
D. Meisel[3], C.D. Jonah[3], and M.C. Sauer, Jr.[3]

(1)   Pacific Northwest Laboratory
        Richland, Washington 99352
(2)   Georgia Institute of Technology
        Atlanta, Georgia
(3)   Argonne National Laboratory
        Chicago, Illinois

INTRODUCTION

In the early 1940s, production of nuclear materials for use in weapons began at the newly established Hanford works under the Manhattan Project. Although several chemical processes were used to separate plutonium from uranium and fission products, all of the wastes were stored in underground mild steel tanks ranging in size from about 190 m^3 to 3800 m^3. Use of mild steel required that the wastes be made alkaline before being placed in the tanks.

During the 1950s, recovery of valuable uranium from the tanks led to projected increases in waste volume that exceeded the available tank space. To decrease the volume of the stored waste, a ferrocyanide scavenging process was developed in which $Na_2NiFe(CN)_6$ [ideal stoichiometry] was precipitated in mildly alkaline solution. This precipitate carried with it most of the ^{137}Cs. Changes to the process also allowed for the precipitation of some of the ^{90}Sr and ^{60}Co. The decontaminated solutions were then sent to soil cribs for final disposal. The solids and some liquids remained in the tanks.

As various other chemical processes were developed, the volume of the waste continued to increase. Eventually, the decision was made to transfer all liquids from the single-shell tanks used in the early years to new double-shell tanks. Although other techniques were tested for reducing the liquid volume before transfer to the double-shell tanks, the bulk of the liquid both from the then operating plants and the liquid recovered from the single-shell tanks were passed through crystallizer-evaporator units. The volume of the each double-shell tank exceeds 3800 m^3 and the 28 tanks contain a total volume of $2.3 \cdot 10^5$ m^3 of stored waste (Hanlon 1993).

Over the years, waste transfers into and out of the tanks led to mixing of waste types, and to the resulting safety concerns. These safety concerns fall into six categories:

Chemical Pretreatment of Nuclear Waste for Disposal, Edited by
W.W. Schulz and E.P. Horwitz, Plenum Press, New York, 1995

1. wastes that generate flammable gases or gas mixtures

2. wastes that contain high concentrations of ferrocyanides or tanks suspected of containing large amounts of ferrocyanides

3. wastes that contain greater than 3 wt% total organic carbon

4. wastes from which toxic or noxious vapors are suspected of emanating

5. wastes that contain high radiolytic heat

6. wastes that may contain sufficient fissile material to pose a criticality concern.

Of these categories, only the chemistry associated with the generation of flammable gases is discussed in this paper.

HISTORY OF TANK 241-SY-101

Of 24 tanks, both single- and double-shell, that are suspected of containing flammable gas generating wastes, the waste in Tank 241-SY-101 stands out as unique and has been most studied. Tank 241-SY-101 is most notable because the episodic release of flammable gases is the most spectacular. Waste was first added to this tank in 1977 and was also added during several campaigns from 1977 to 1981. The fill history has been documented (Strachan et al. 1990), and the major components of the waste are shown in Table 1 (Herting et al. 1992).

As the last waste was added to the tank, the waste volume began to increase, and a short time later the waste volume rapidly decreased. This episodic growth and release is illustrated in the data shown in Figure 1 and continues to this day. The periodicity has, until recently, been fairly regular at 90 to 110 days between decreases. Recently, the period has lengthened to about 150 days.

Studies done in the early 1980s (Delegard 1980) with simulated wastes in the absence of radiation revealed that the growth of the slurry was due to the generation of gases. These gases were found to consist of H_2, N_2O, N_2, and possibly NH_3. In 1990, a concerted effort was begun to study the behavior of the waste in this tank. Gas from the tank was indeed found to consist of the same gases identified by Delegard (1980). More disconcerting was the discovery that the concentration of H_2 was often found to exceed the lower flammability limit (4%) during these gas release events and on

Table 1. Molar Composition of Waste in Tank 241-SY-101 - Major Constituents

Component	Bulk	Liquid
$NaNO_3$	2.2	2.7
$NaNO_2$	4.0	3.8
$NaAlO_2$	2.0	1.7
Na_2CO_3	0.4	0.1
NaOH	2.4	2.3
Total Organic Carbon	2.4	1.2

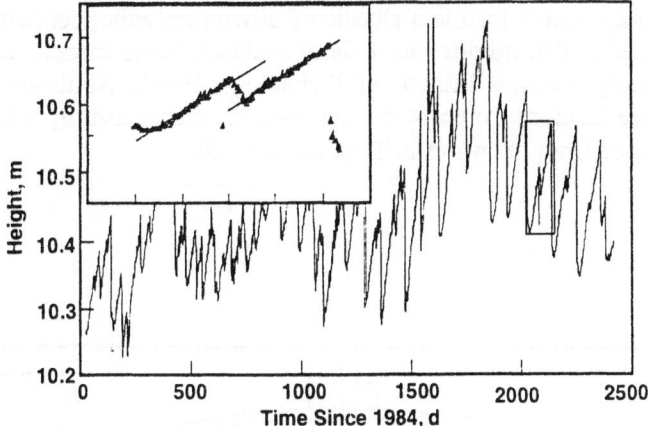

Figure 1. Waste Height Data From Tank 241-SY-101.

occasion to exceed 5% for a few minutes.[1] These observations led to much safer operations and to intensive studies to determine the chemistry and physics of gas generation and release in Tank 241-SY-101.

Recent investigations have been summarized in several documents (Strachan and Morgan 1990, Strachan and Morgan 1991 a, b, c; Strachan 1992 a, b; Schulz and Strachan 1992; Babad et al. 1992 and others). From a few weeks after a gas release event until the next release event, the waste appears to consist of four layers as depicted in Figure 2. A sludge is thought to exist at the bottom of the tank[2]. The presence of a sludge layer is surmised from the thermal behavior in this region during a gas release event; the temperature does not change as it does elsewhere. Above the sludge is a layer of slurry approximately 4.7 m (180 in.) thick. This layer is thought to be nonconvecting because during the period between gas release events, a nearly parabolic temperature profile is developed. Above this layer is thought to lie a layer that is also about 4.7 m thick. This layer is thought to be convecting because of the nearly isothermal profile most of the time. The uppermost layer is thought to be a crust or foam-like material. In this layer, the temperature decreases nearly linearly from the temperature of the bottom of the layer to the temperature of the dome space. Further discussion of the physics of the waste in this tank is beyond the scope of this paper. Therefore, the remainder of the discussion centers on the chemistry of the waste in Tanks 241-SY-101.

RESULTS FROM CHEMICAL MECHANISM STUDIES

It is necessary to understand the mechanism by which gases are generated in Tank 241-SY-101 because the waste will have to be treated at some time for final disposal - the wastes cannot be stored indefinitely in these steel tanks. Whatever disposal method is selected must deal with the complex chemistry of these wastes so that the treatment

---

(1)   The tank head space is ventilated using an exhaust fan at 0.2 to 0.3 m^3/s. The volume above the waste in the tank is about 140 m^3.

(2)   This layer may no longer exist in the tank because of the vigorous stirring action of some episodic gas releases.

leads to safe storage rather than to a situation that requires either constant monitoring or further treatment. Toward this end, a set of studies is being carried out at Argonne National Laboratory, Georgia Institute of Technology, Pacific Northwest Laboratory, and Westinghouse Hanford Company to determine the mechanism by which the gases are being generated from the waste in Tank 241-SY-101.

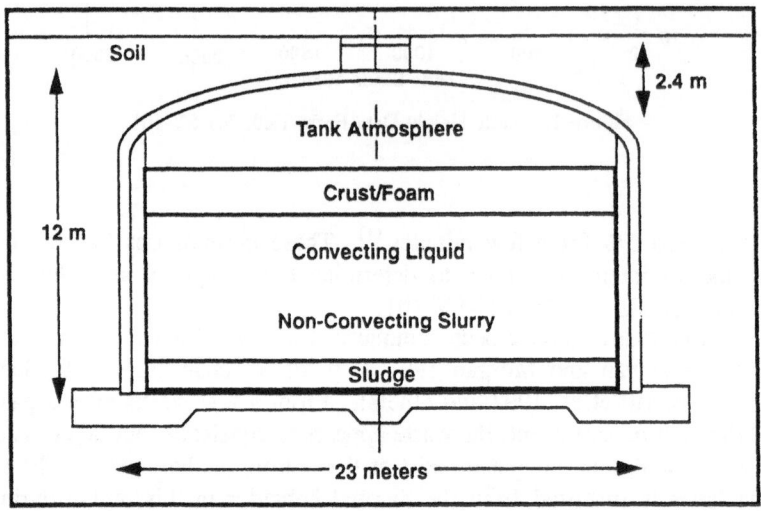

Figure 2. Schematic of Tank 101-SY.

Mechanistic studies are being carried out at the Georgia Institute of Technology using simulated waste materials in the absence of radiation. The composition of this waste is based on the inorganic analyses of the actual waste (Table 1) and an educated guess about the organic constituents. This educated guess is based on process history, and also on the work by Delegard (1980 and 1987) and Lokken et al. (1986). It was found that a dilute solution of formaldehyde in strong base (Siemer in Strachan 1991, Ashby et al 1993) yielded hydrogen and not the typical Cannizzaro reaction products. Delegard (1980) determined that the presence of aluminum, presumably as $Al(OH)_4^-$, was necessary for the reaction to produce H_2. Similarly, nitrite and base were also needed for the reaction. From this information and some preliminary work with ^{13}C-labelled and ^{15}N-labelled organic compounds, the prototypic reactions shown in Figure 3 for the thermal degradation of glycolate were proposed (see Ashby et al. in Strachan 1992b or Schulz and Strachan 1992).

The validity of most of these reactions has been confirmed from studies with ^{13}C-labelled organic chemicals (Ashby et al. in Strachan 1992a,b; in Schulz and Strachan 1992). The observation that only NO_2^- participates in the reactions is supported by the studies with ^{15}N-labelled NO_2^- and/or NO_3^-. The NO_2^- reacts in the presence of the organics to yield N_2O and N_2; NH_3 appears to be a product of the organic nitrogen, but not entirely. This information is summarized in Tables 2 and 3.

(1)	$Al(OH)_4^- + NO_2^- \rightleftarrows Al(OH)_3\text{-}O\text{-}N\text{-}O^- + OH^-$
(2)	$Al(OH)_3\text{-}O\text{-}N=O + HO\text{-}CH_2\text{-}CO_2^- \rightleftarrows Al(OH)_4^- + O=N\text{-}O\text{-}CH_2\text{-}CO_2^-$
(3)	$O=N\text{-}O\text{-}CH_2\text{-}CO_2^- \rightarrow NO^- + O=CH_2 + CO_2$
(4)	$O=N\text{-}O\text{-}CH_2\text{-}CO_2^- + OH^- \rightarrow NO^- + H\text{-}(CO)\text{-}CO_2^- + H_2O$
(5)	$2NO^- \rightleftarrows N_2O_2^{2-}$
(6)	$N_2O_2^{2-} + H_2O \rightleftarrows HN_2O_2^- + OH^-$
(7)	$HN_2O_2^- \rightarrow N_2O + OH^-$
(8)	$N_2O + Al(OH)_3\text{-}O\text{-}N\,O^- \rightarrow N_2 + Al(OH)_3\text{-}O\text{-}NO_2$
(9)	$N_2O + NO_2^- \rightarrow N_2 + NO_3^-$
(10)	$CH_2=O + OH^- \rightleftarrows HO\text{-}CH_2\text{-}O^-$
(11)	$HO\text{-}CH_2\text{-}O^- + OH^- \rightleftarrows {}^-O\text{-}CH_2\text{-}O^- + H_2O$
(12)	${}^-O\text{-}CH_2\text{-}O^- + H_2O \rightarrow H_2 + H\text{-}COO^- + OH^-$
(13)	$HC(O)CO_2^- + OH^- \rightleftarrows {}^-O\text{-}CH(OH)\text{-}CO_2^- \rightleftarrows ({}^-O)_2\text{-}CH\text{-}CO_2^-$
(14)	$({}^-O)_2\text{-}CH\text{-}CO_2^- + H_2O \rightarrow H_2 + {}^-O_2C\text{-}CO_2^- + OH^-$

Figure 3. Proposed Mechanism of Thermal Degradation of Glycolate Ion.

Table 2. Origin of Gases Generated Under Thermal Conditions

Component	Gas
CH_2O	H_2
Glycolate	H_2
NO_2^-	N_2O, N_2, NH_3[a]
R_3N[b]	N_2O, N_2, NH_3[a]

(a) about 90% of the NH_3 originates from the NO_2^- and 10% from the organic amine.

(b) R = H and/or alkyl group

Table 3. Summary of the Results from the Studies with ^{13}C-Labelled Organics

Complexant	Products
HEDTA[1]	CH_2O, $C_2O_4^{2-}$, EAMA, EA, [13,12]C-glycine
HEDTA[2]	HCO_2^-, CO_2, $C_2O_4^{2-}$, EAMA, u-EDDA, IDA, s-EDDA or EDMA, ^{13}C-glycine
Glycolate	$C_2O_4^{2-}$

1   [13]C on the β-hydroxyethyl group
2   [13]C on the β-carboxymethyl groups
HEDTA - (2-hydroxyethylene) ethylenediaminetriacetate
EAMA - ethanolaminemonoacetate
EA - ethanolamine
EDDA - ethylenediaminediacetate (u- unsymmetrical, s - symmetrical)
IDA - iminodiacetate
EDMA - ethylenediaminemonoacetate

In the suggested thermal reaction mechanism, two-carbon moieties are removed from the organic molecule and, depending on the electron rearrangement, a two- or single-carbon moiety eventually yields H_2. The case of glycolate is, of course, much simpler, but provides the basic understanding of the decomposition of the more complex molecules.

While these experiments have elucidated what may be the foundation for the reactions that take place in the actual waste, the gas compositions that result from these and other laboratory experiments, including the radiolysis experiments described below, are much different from the gas compositions that are observed in the tank (Pederson and Strachan 1993). The exact reason for this difference is uncertain at this time, but may be due to the difference between the organic composition of the waste in the tank and the compositions used in the laboratory studies. The organics used in the laboratory studies were selected based upon a knowledge of the chemical processes that took place at Hanford. The actual organics in the tank are largely unknown. The fact that the gas composition is a sensitive function of the simulated waste composition, including trace elements such as transition metals, noble metals, and Cl^-, has been documented by Bryan and Pederson (in Strachan 1992b; Schulz and Strachan 1992).

Reactions in the presence of radiation are being carried out at Argonne National Laboratory. In these studies, gas is measured using sensitive devices that have been developed over many years of studying radiation chemistry. These techniques are described by Meisel (in Strachan et al. 1991a,b and Meisel et al. 1991a,b). From these studies, it has been shown that the amount of gas generated from the radiolysis of water is surpassed by the amount generated from the radiolysis of simulated waste solutions. The direct generation of gases by radiolysis constitutes about 20% of the total gas generated during these laboratory experiments. It appears that the presence of radiation causes other reactions to occur that would not occur without the radiation. These reactions are the result of free radicals that are generated from the radiation. Although the radiation chemistry of the $OH^-/NO_3^-/NO_2^-$ system is well known (Meisel et al. 1991a,b and in Strachan 1992a), the contribution to the radiation chemistry by the organics is not well understood.

It has been found that the $G(H_2)$[1] is independent of dose and dose rate. This is illustrated in Figure 4 for a simulated waste solution containing EDTA. For N_2O, the $G(N_2O)$ has a limited dependence on dose rate; at lower dose rates, the $G(N_2O)$ increases. This may be due to the conversion of N_2O to N_2 and O_2 in the radiation field. As in the studies of the thermal generation of gases, it has been determined that the source of the N_2O is NO_2^-, not the organic nitrogen nor the NO_3^-.

The case for free radical generation or reactive intermediates is made stronger by the observation that preirradiation of the simulated waste solutions before performing thermal gas generation experiments leads to an increase in the gas generation rate (Figure 4). Solutions that had been set aside after determining the $G(H_2)$ and $G(N_2O)$ were heated later to determine the rate of thermal generation of gases. Additional experiments were performed, and the results are shown in Table 4.

From these data it can be seen that the temperature sensitivity of the N_2O production is greater than the H_2 production. Although the "Control" and the "Preirradiated" solutions contained both EDTA and HEDTA in equimolar concentrations, it has been shown that these reactions are essentially independent, i.e. the effects of each are additive (see Meisel in Strachan 1992b). As with the results from the thermal experiments described above and those that have been performed at Pacific Northwest

_____

(1) The "G-value" is the number of molecules generated, in this case H_2, per 16 aJ (100 eV) of radiation energy.

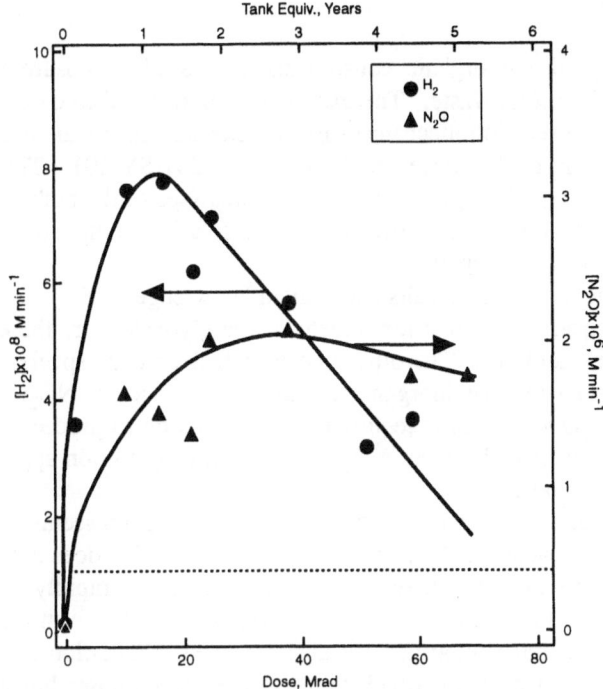

Figure 4. Effect of Preirradiation on the rate of thermal generation of H_2 and N_2O for the Solution Shown in the Footnote of Table 4 with 0.17 m IDA.

Table 4. Results for $G(H_2)$ and $G(N_2O)$ from the Preirradiation Experiments.

Sample	Concentration M	$G(H_2)$ 60°C	30°C	Ratio	$G(N_2O)$ 60°C	30°C	Ratio
EDTA	0.085	0.062	0.043	1.44	1.10	0.54	2.04
HEDTA	0.085	0.054	0.047	1.15	0.77	0.54	1.43
Control	0.065	0.063	0.045	1.40	0.87	0.49	1.78
Preirrad	0.065	0.080	0.047	1.70	1.06	0.48	2.21

Note: Both the Control and the Preirradiated solutions contained 0.065 M each of EDTA and HEDTA; the total dose was 130 krad at 4.4 krad/min; the ratio is the 60°C value divided by the 30°C value; the solution used had an inorganic composition of 2.12 M NaOH, 2.79 M NaNO$_3$, 2.22 M NaNO$_2$, and 1.30 M NaAlO$_2$.

Laboratory and Westinghouse Hanford Company, the generation rates and the relative concentrations of the gases produced in the laboratory experiments do not match the values observed for the gases in the tank vapor space.

In general, N_2O is generated in laboratory experiments 5-to-10 times faster than H_2, whereas in the tank these two gases appear to be generated at nearly equal rates. In laboratory experiments, NH_3 and N_2 are generated at much slower rates than H_2. In the tank the generation rate of NH_3 is about 2 to 3 times that of H_2, and N_2 generation cannot be measured with accuracy.[1]

---

(1)   These reactions (#8 and #9 in Figure 3), while consistent with the observed chemistry, have since been shown not to occur.

CONCLUSIONS

The identity of the organic constituents is needed to ensure that the waste simulants mimic the actual waste. The experiments performed to date have elucidated large portions of the reaction mechanism for the chelator chemicals that make up part of the organic content of the waste solutions in Tank 241-SY-101. These mechanisms must await confirmation through accurate organic analyses of the tank waste. Analyses are ongoing, but are complicated because of the solution complexity and the highly radioactive nature of the samples.

Based on the laboratory results and limited knowledge of the organic constituents in the waste, it appears likely that formaldehyde and glyoxalate are the key species that react with base to yield H_2. The other two principal gaseous species (N_2O and N_2) appear to originate from the inorganic chemicals, specifically NO_2^-. However, the presence of the organic species is required for this reaction to proceed. The source of NH_3 appears to be principally from NO_2^-, but a significant portion appears to originate from the organic amines.

While the thermal generation of gases from the chelators seems well understood, radiation affects the reactions. In the thermal reactions, EDTA does not react very fast. However, in the presence of radiation, the reaction proceeds rapidly. Therefore, it is expected that the products in the actual waste may be different from those observed in the thermal reaction, but similarities are also expected. Radiation is not expected to change the basic mechanism by which these species decompose; however, additional products are present from the free radicals produced by radiation. Subsequent reaction of these products would be expected to follow routes similar to those shown in Figure 3.

Progress is being made in the understanding of both the physical and chemical behavior of the waste in Tank 241-SY-101. As this understanding evolves, a better approach toward the mitigation of the potential safety hazard associated with this waste can be identified.

REFERENCES

Ashby, E. C., Doctorovitch, F., Liotta, C. L., Neumann, H. M., Barefield, E. K., Konda, A., Zhang, K., Hurley, J., and Siemer, D., 1993, "Concerning the Formation of Hydrogen in Nuclear Waste: Quantitative Generation of Hydrogen via a Cannizzaro Intermediate," *Journal of the American Chemical Society*, 115:1171.

Babad, H., Johnson, G. D., Lechelt, J. A., Reynolds, D. A., Pederson, L. R., Strachan, D. M., Meisel, D., Jonah, C., and Ashby, E. C., 1991, "Evaluation of the Generation and Release of Flammable Gases in Tank 241-SY-101," WHC-EP-0517, Westinghouse Hanford Company, Richland, Washington.

Delegard, C. H., 1980, "Laboratory Studies of Complexed Waste Slurry Volume Growth in Tank 241-SY-101," RHO-LD-124, Rockwell Hanford Operations, Richland, Washington.

Delegard, C. H., 1987, "Identities of HEDTA and Glycolate Degradation Products in Simulated Hanford High-Level Waste," RHO-RE-ST-55P, Rockwell Hanford Operations, Richland, Washington.

Hanlon, B. M., 1993, "Tank Farm Surveillance and Waste Status Summary Report for October 1992," WHC-EP-0182-55, Westinghouse Hanford Company, Richland, Washington.

Herting, D. L., Bechtold, D. B., Crawford, B. A., Welsh, T. L., and Jensen, L., 1992, "Laboratory Characterization of Samples Taken in May 1991 from Hanford Waste Tank 241-SY-101," WHC-SD-WM-DTR-024 (Revision 0), Westinghouse Hanford Company, Richland, Washington.

Lokken, R. O., Scheele, R. D., Strachan, D. M., and Toste, A. P., 1986, "Complex Concentrate Pretreatment FY 1986 Progress Report", PNL-7687, Pacific Northwest Laboratory, Richland, Washington.

Meisel, D., Diamond, H., Horwitz, E. P., Jonah, C. D., Matheson, M. S., Sauer, M. C., Jr., Sullivan, J. C., Barnabas, F., Cerny, E., and Cheng, Y. D., 1991a, "Radiolytic Generation of Gases from Synthetic Waste. Annual Report - 1991," ANL-91/41, Argonne National Laboratory, Argonne, Illinois.

Meisel, D., Diamond, H., Horwitz, E. P., Jonah, C. D., Matheson, M. S., Sauer, M. C., Jr., and Sullivan, J. C., 1991b, "Radiation Chemistry of Synthetic Waste," ANL-91/40, Argonne National Laboratory, Argonne, Illinois.

Pederson, L. R., and Strachan, D. M., 1993, "Status and Integration of the Gas Generation Studies Performed for the Hydrogen Safety Program - FY-1992 Annual Report," PNL-8523, Pacific Northwest Laboratory, Richland, Washington.

Schulz, W. W., and Strachan, D. M., 1992, "Minutes of the Tank Waste Science Panel Meeting March 25-27, 1992," PNL-8278, Pacific Northwest Laboratory, Richland, Washington.

Strachan, D. M., 1991, "Minutes of the Tank Waste Science Panel Meeting February 7-8, 1991," PNL-7709, Pacific Northwest Laboratory, Richland, Washington.

Strachan, D. M., 1992a, "Minutes of the Tank Waste Science Panel Meeting November 11-13, 1991," PNL-8047, Pacific Northwest Laboratory, Richland, Washington.

Strachan, D. M., 1992b, "Minutes of the Tank Waste Science Panel Meeting July 9-11, 1991," PNL-8048, Pacific Northwest Laboratory, Richland, Washington.

Strachan, D. M., and Morgan, L. G., 1991a, "Minutes of the Tank Waste Science Panel Meeting September 13-14, 1990," PNL-7599, Pacific Northwest Laboratory, Richland, Washington.

Strachan, D. M., and Morgan, L. G., 1991b, "Minutes of the Tank Waste Science Panel Meeting July 20, 1990," PNL-7598, Pacific Northwest Laboratory, Richland, Washington.

Strachan, D. M., and Morgan, L. G., 1991c, "Minutes of the Tank Waste Science Panel Meeting June 26-27, 1990," PNL-7602, Pacific Northwest Laboratory, Richland, Washington.

COMBINED TRUEX-SREX EXTRACTION/RECOVERY PROCESS

E. Philip Horwitz, Mark L. Dietz, Herbert Diamond, and
Robin D. Rogers

Chemistry Division, Argonne National Laboratory
Argonne, IL 60439-4831 USA

and

Ralph A. Leonard

Chemical Technology Division, Argonne National Laboratory
Argonne, IL 60439-4837 USA

INTRODUCTION

The United States Department of Energy Hanford site defense complex has the largest volume of stored nuclear waste in the US. The complex has been in existence for nearly fifty years, and the varied chemical operations over this time period have generated a wide variety of wastes. The aqueous wastes, both high-level and TRU (transuranic), are stored in underground single-shell and double-shell tanks. Because the aqueous wastes are made alkaline (pH ~14) to prevent corrosion of the carbon steel liners of the tanks, the uranium, TRUs and ^{90}Sr, together with most of the fission products, are largely concentrated in the insoluble sludge formed by the hydroxides, phosphates, and silicates of metal ions such as aluminum, iron, zirconium, and bismuth. Fission products cesium and technetium are largely concentrated in the supernatant solution.

Plans are currently being formulated to dispose of this defense waste. There is a strong economic incentive to pretreat the stored waste to minimize the quantity of waste requiring vitrification and subsequent burial in a deep geologic repository. Because of the diversity and complexity of the waste stored in single- and double-shell tanks, the demands on any separation process are great. Perhaps the greatest difficulty for the separations chemist is the development of processes that can remove a number of the hazardous radionuclides in a single operation. The objective of this program is to develop a process in which strontium and tectnetium, as well as uranium and TRUs, are extracted from the dissolved sludge from Hanford waste storage tanks. A further objective is to selectively

Chemical Pretreatment of Nuclear Waste for Disposal, Edited by
W.W. Schulz and E.P. Horwitz, Plenum Press, New York, 1995

81

partition strontium and TRUs from uranium and tectnetium during the stripping operation. Uranium and technetium would then be stripped from the process solvent and separated from each other by a tectnetium selective chromatographic material. Uranium would eventually be routed to storage and strontium, tectnetium, and TRUs vitrified for burial. Such a process would result in a significant reduction in the amount of waste requiring vitrification because the raffinate from the process can be converted to grout, although it may still require removal of ^{137}Cs to be classified as low-level waste.

The basic problem in designing solvent extraction processes to remove such a wide range of elements is that no single ligand is capable of coordinating selectively to elements with such widely differing charge densities and hydration energies. However, a possible solution to multi-element extraction is to mix two extractants with totally different chemical properties into a single process solvent formulation. For this approach to be successful, both extractants must be soluble in the same diluent and show no strong interaction.

Earlier studies (Horwitz, 1992) have shown the feasibility of combining a transuranic (TRU) selective extractant (Horwitz and Schulz, 1991), octyl(phenyl)-N, N-diisobutyl-carbamoylmethylphosphine oxide (CMPO), and a strontium selective extractant (Horwitz, et al., 1991), 4, 4' (5') bis-tertiarybutylcyclohexano-18-crown-6 (DtBuCH18C6 or CE), into a single process solvent formulation with the aid of a phase modifier, tri-n-butyl phosphate (TBP). Both CMPO and DtBuCH18C6 (CE) are neutral extractants. Figures 1 and 2 show that when dissolved in TBP, CMPO and CE have the same efficiency for americium and strontium extraction, respectively (as measured by the distribution ration, D_{Am} and D_{Sr}), whether present individually or in a mixture. The data in Figures 1 and 2 show that strontium and americium (and presumably tetravalent and hexavalent actinides) can be extracted by a single process solvent when CMPO and DtBuCH18C6 are mixed.

This report describes our attempts to develop a combined transuranic-strontium extraction/recovery process based on the principle of combining CMPO and DtBuCH18C6 into a single process solvent formulation. We refer to this process hereafter as the Combined TRUEX-SREX Extraction/Recovery Process or simply the Combined Process.

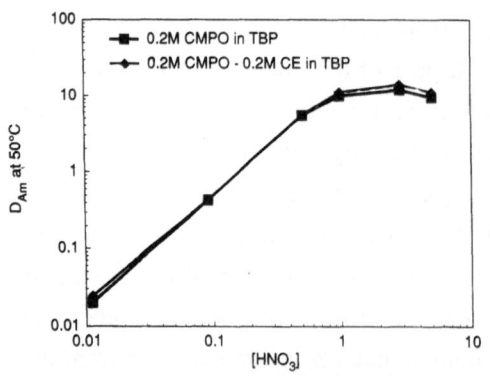

Figure 1. Acid dependency of D_{Am}. ◊ - 0.2 M CMPO in TBP, ◻ - 0.2 M CMPO - 0.2 M DtBuCH18C6 in TBP. T = 50 °C.

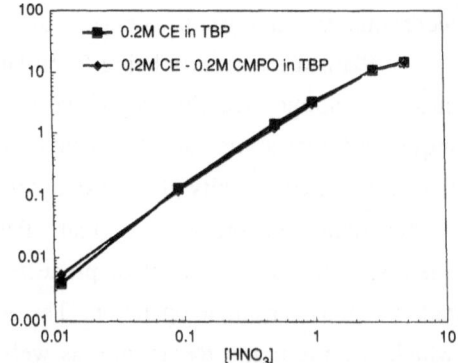

Figure 2. Acid dependency of D_{Sr}, ◊ - 0.2 M DtBuCH18C6 in TBP, ◻ - 0.2 M DtBuCH18C6 -0.2 M CMPO in TBP. T = 50 °C.

The Combined Process should reduce the cost of chemical pretreatment because it would require less space and equipment and would be easier to control than a TRUEX and SREX process operated in tandem.

EXPERIMENTAL

Reagents

Di-t-butylcyclohexano-18-crown-6 (DtBuCH18C6) was obtained from Parish Chemical Co. (Vineyard, UT) and used without further purification. Octyl(phenyl)-N,N-diisobutyl-carbamoylmethylphosphine oxide (CMPO) was obtained from Atochem North America and recrystallized from hexane as described in (Gatrone et al., 1987). Tri-n-butyl phosphate (TBP) was obtained from Eastman Chemical Co. and distilled before use. Diamyl amylphosphonate (DA[AP]) was obtained from Eichrom Industries, Inc. (Darien, IL) and used as received. Dodecane was obtained from Aldrich Chemical Co. and Norpar™-12 and Isopar™-L and -M were obtained from Exxon Company. Ultrex™-grade nitric acid (J.T. Baker Chemical Co.) and Milli -Q2 purified water were used to prepare all nitric acid solutions.

Distribution Ratios and Third Phase Formation

Distribution ratios of strontium, tectnetium, bismuth, and americium and third phase formation were measured using the procedures described previously (Horwitz et al. 1982). Isotopes used were ^{85}Sr, ^{99}Tc, ^{207}Bi, ^{241}Am, ^{239}Pu, ^{233}U, and natural uranium. All organic phases were preconditioned by 2-3 contacts with an aqueous phase containing the appropriate HNO_3 concentration. A one minute mixing time was used throughout. Standard radiometric assay and counting procedures were employed. Duplicate assays of each phase were routinely performed. Radiometrically measured distribution ratios have a standard deviation of ±5%.

Physical Properties of Process Solvents

In preparation for laboratory tests of the combined TRUEX-SREX process using a 20-stage minicontactor, three solvent properties were measured: density, viscosity, and dispersion number (N_{Di}). Density was determined by weighing the solvent in a volumetric flask previously calibrated using distilled water. Viscosity data were obtained using a Brookfield Viscometer (Model LVF) with an ultra-low adapter. N_{Di} was evaluated by measuring the time-to-break of the two phases, the solvent and the aqueous phase, after they were dispersed by shaking in a 100-mL graduated cylinder. The N_{Di} was calculated from the following equation:

$$N_{Di} = \frac{1}{t_B} \sqrt{\frac{\Delta Z}{g}}$$

where t_B is the time in seconds required for the phases to disengage, ΔZ is the total height of the two phases in meters, and g is the gravitational constant, 9.81 m/s^2 (Leonard et al. 1982).

RESULTS AND DISCUSSION

DtBuCH18C6 and CMPO in TBP

Although the data in Figures 1 and 2 show that the simultaneous extraction of strontium and americium is feasible by mixing DtBuCH18C6 and CMPO with TBP, the data do not show how D_{Sr} and D_{Am} are affected by the concentration of the extractants. Tables 1 and 2 show the dependency of D_{Sr} and D_{Am}, respectively, on extractant concentration. (All measurements using undiluted TBP were carried out at 50 °C to decrease the viscosity of the organic phase.) The first power dependency of D_{Sr} on crown ether concentration is expected based on previous reports on strontium extraction using the DtBuCH18C6 macrocycle (Horwitz et al., 1991). Also expected is the change in extractant dependency of D_{Am} from second power to third power as the ratio of CMPO/TBP ratio increases (Chiarizia and Horwitz, 1987). Tables 3 and 4 show the nitric acid dependency of D_{Sr} and D_{Am}, respectively, for 0.1, 0.2, and 0.3 $\underline{M}$ extractant. These three extractant concentrations cover the most likely concentrations that would be employed in the combined solvent formulation. The data in Tables 1 through 4 show that 0.2 $\underline{M}$ is an effective concentration for each extractant in the combined solvent formulation. Higher concentrations of CMPO will increase the difficulty of stripping americium, while higher concentrations of either DtBuCH18C6 or CMPO will increase the viscosity of the process solvent, even when a diluent/phase modifier combination is employed. (See section on physical properties.) Concentrations of extractant below 0.2 $\underline{M}$ would significantly reduce the overall efficiency of process because of the low values of D_{Am} and D_{Sr} in the extraction stages. Figure 3 compares the distribution ratios of strontium, americium, tectnetium, and bismuth as a function of the nitric acid concentration for 0.2 $\underline{M}$ DtBuCH18C6 - 0.2 $\underline{M}$ CMPO-TBP. The data for tectnetium are included because its removal from high-level liquid waste is also desirable. The data for bismuth are included because it is the most extractable of the inert constituents present in the waste. High concentrations of bismuth in the organic phase have

Table 1. Extractant dependency of D_{Sr} for DtBuCH18C6 in TBP Aqueous phase = 1.0 $\underline{M}$ HNO3, T = 50° C

DtBuCH18C6, M	D_{Sr}
0.01	1.7×10^{-1}
0.05	8.9×10^{-1}
0.10	1.5
0.20	3.0
0.30	4.5
0.50	6.9

Table 2. Extractant dependency of D_{Am} for CMPO in TBP Aqueous phase = 1.0 $\underline{M}$ HNO3, T = 50 °C

CMPO, M	D_{Am}
0.01	2×10^{-2}
0.05	7.1×10^{-1}
0.10	2.5
0.20	11
0.30	26
0.50	94

Table 3. Acid dependency of D_{Sr} using DtBuCH18C6 in TBP. T = 50 °C

[HNO3], $\underline{M}$	D_{Sr}		
	DtBuCH18C6 Concentration		
	0.1 M	0.2 M	0.3 M
0.0112	2.3×10^{-3}	5.2×10^{-3}	6.0×10^{-3}
0.0919	6.5×10^{-2}	1.2×10^{-1}	1.5×10^{-1}
0.505	6.3×10^{-1}	1.2	1.7
0.990	1.5	3.0	4.5
2.87	5.3	11	15
5.40	6.9	14	21

Table 4. Acid dependency of D_{Am} using CMPO in TBP. T = 50 °C

[HNO3], $\underline{M}$	D_{Am}		
	CMPO Concentration		
	0.1 M	0.2 M	0.3 M
0.0112	7.8×10^{-3}	2.4×10^{-2}	4.0×10^{-2}
0.0919	9.6×10^{-2}	4.3×10^{-1}	1.1
0.505	1.3	5.4	14
0.990	2.5	11	26
2.87	3.5	14	31
5.40	2.9	11	24

a significant influence on D_{Am}. (See section on chemical properties of the combined solvent.). The data in Figure 3 show that strontium and americium are effectively extracted from >1 $\underline{M}$ HNO₃. Since other actinides such as Th, uranium, neptunium, and plutonium are even more strongly extracted than americium, the mixture of DtBuCH18C6 and CMPO in the presence of a phase modifier shows considerable promise as a combined TRUEX-SREX process solvent.

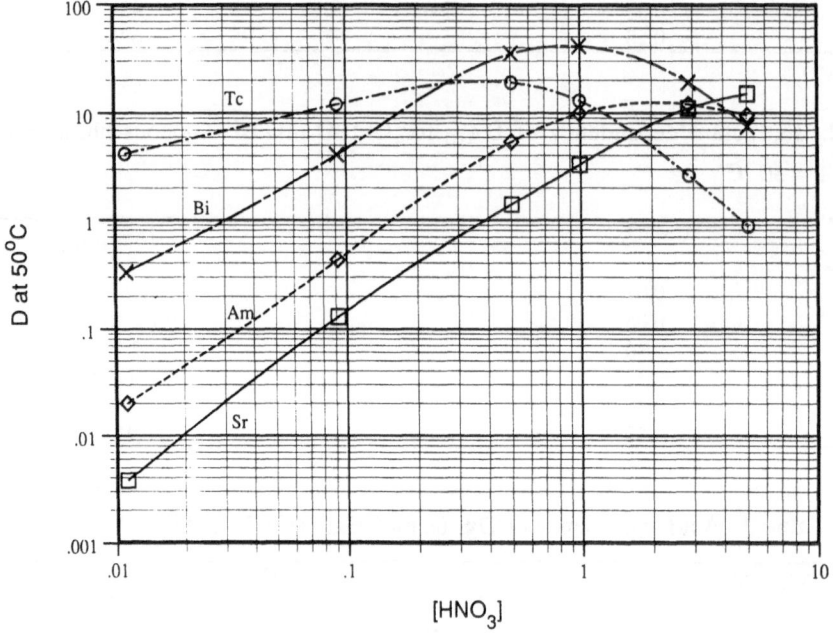

Figure 3. Distribution ratio of Am, Sr, Tc, and Bi vs HNO₃ for 0.2 $\underline{M}$ DtBuCH18C6 - 0.2 $\underline{M}$ CMPO in TBP. T = 50 °C.

Combined Process Solvent: Physical Properties

A major problem that frequently arises when extractants are mixed is the formation of a second heavy organic phase (i.e., third phase formation) when the solvent is loaded with metal ions. Both CMPO and DtBuCH18C6 require the presence of a phase modifier, e.g., TBP, if a hydrocarbon is used as the diluent. Table 5 shows the influence of the carbon chain length of the diluent on the loading of uranyl nitrate in a 0.2 $\underline{M}$ CMPO - 0.2 $\underline{M}$ DtBuCH18C6 solution. Two different phase modifiers were used, TBP and DA[AP], with three different diluents. (Norpar™-12 is a normal paraffinic hydrocarbon whereas Isopar™-L and -M are isoparaffinic hydrocarbons.) The data in Table 5 show the maximum concentration of uranyl nitrate that can be extracted into the organic solvent mixture without the formation of a third phase. Uranyl nitrate was chosen to induce third phase formation because uranium, together with bismuth, are the major extractable constituent in the sludge from the Hanford storage tanks.

Table 5. Influence of diluent and phase modifier on loading of process solvent formulations* (Aqueous Phase 3 $\underline{M}$ HNO₃ - Uranyl Nitrate, 25 °C)

Phase Modifier-Diluent	No. of Carbons	Flash Point	[U]org, M**
TBP-Dodecane	12.0	71	0.021
TBP-Norpar™ - 12	11.5	69	0.038
DA[AP]-Norpar™ - 12	11.5	69	0.11
TBP-Isopar™ - L	12.0	61	0.14
DA[AP]-Isopar™ - L	12.0	61	>0.20
TBP-Isopar™ - M	13.5	80	0.029
DA[AP]-Isopar™ - M	13.5	80	0.20

* Process solvent 0.2 $\underline{M}$ CMPO-0.2 $\underline{M}$ DtBuCH18C6-1.2 $\underline{M}$ phase modifier-diluent.
** Maximum concentration of uranium in the process solvent without third phase formation.

The data show that to minimize third phase formation while simultaneously utilizing a diluent with a high flash point (for safety), isoparaffinic hydrocarbons are more effective than normal paraffinic hydrocarbons. However, 1.2 $\underline{M}$ TBP in Isopar™-M forms a third phase above 0.029 $\underline{M}$ [U]$_{org}$, which significantly reduces the efficiency of the extraction stages because a high ratio of organic to aqueous phase (O/A) will be required to maintain [U]$_{org}$ below 0.03 $\underline{M}$ concentration. As can be seen in Table 5, substitution of DA[AP] for TBP significantly reduces third phase formation with all diluents. The superiority of neutral phosphonate esters over TBP as phase modifiers was observed in an earlier study (Kolarik and Horwitz, 1988). From the above data, the two most promising combinations of phase modifier and diluent for the process solvent are DA[AP]-, -Isopar™-L and -Isopar™-M.

Although high flash point and minimization of third phase formation during metal ion loading are very important properties of a process solvent, the viscosity and density of the process solvent are also important. Process solvents with high viscosities and densities close to those of the aqueous phase disengage slowly when mixed with aqueous phases. The data in Table 6 compare the viscosity and density for the DtBuCH18C6-CMPO-DA[AP] combination in Norpar™-12, -Isopar™-L and Isopar™-M. Uranyl nitrate loading data from 5 $\underline{M}$ HNO₃ are also included for comparison. (Note that [U]$_{org}$ data shown in Table 5 were obtained from 3 $\underline{M}$ HNO₃. Generally, the tendency towards third phase formation increases with increasing HNO₃ concentration. Therefore, the [U]$_{org}$ values in Table 6 are lower than those in Table 5, except for Norpar™-12.) The data in Table 6 show that use of Isopar™-M results in a process solvent with the highest density and viscosity of the three diluents. The significantly higher viscosity of process solvent formulated with Isopar™-M is noteworthy.

Tables 7 and 8 show the dispersion numbers and aqueous entrainment (aqueous in organic phase) for four selected phase modifier-diluent combinations. The dispersion number (N_{Di}) is a measure of how rapidly immiscible phases separate after mixing. The higher N_{Di}, the more rapidly the organic and aqueous phases disengage. The nominal goal is to achieve N_{Di} values of 8×10^{-4} at all O/A phase ratios and to achieve less than 1% aqueous entrainment in the organic phase. The data in Tables 7 and 8 show that these goals are not achieved except in a few cases. Generally N_{Di} decreases and A in O increases as the O/A ratio increases. Therefore, phase disengagement is slower and aqueous entrainment is

Table 6. Comparison of density, viscosity, and uranium loading of the 0.2 $\underline{M}$ CMPO - 0.2 $\underline{M}$ DtBuCH18C6 - 1.2 $\underline{M}$ DA[AP] mixture in different diluents

Phase Modifier-Diluent	Density* (g/ml)	Viscosity* (cps)	[U]**org, $\underline{M}$
Norpar™ - 12	0.855	4.3	0.12
Isopar™ - L	0.865	5.4	0.19
Isopar™ - M	0.875	8.4	0.16

* The densities and viscosities were measured at room temp. (23.5 - 25 °C).
** Maximum concentration of uranium in the process solvent without the formation of a third phase. The organic phase is in equilibrium with 5 $\underline{M}$ HNO_3.

greater during scrubbing and stripping, i.e., high O/A and low acidity. At high acidities, i.e., >1 $\underline{M}$ HNO_3, and O/A <1, N_{Di} values are higher and A in O values are lower than those shown in Tables 7 and 8 and generally meet the nominal goal. For all conditions, the amount of organic phase in the aqueous phase (O in A) is acceptable, that is, O in A is always less than 1%. Of the four phase modifier-diluent combinations shown in Tables 7 and 8, DA[AP]-Isopar™-M is unacceptable whereas the other three are barely acceptable. Of the three borderline acceptable process solvents, the DtBuCH18C6-CMPO-DA[AP]-Isopar™-L combination is considered the most favorable. This selection is based on the assumption that it is easier to overcome the problems associated with the low flash point than with third phase formation.

Table 7. Dispersion numbers*(N_{Di}) and aqueous entrainment (A in O)** for different process solvent formulations. 0.2 $\underline{M}$ CMPO - 0.2 $\underline{M}$ DtBuCH18C6 - Phase modifier-diluent - Room temperature (23.5 to 25.0 °C)

Aqueous phase = 0.1 $\underline{M}$ HNO3

Phase Modifier-Diluent	O/A = 0.33		O/A = 1.0		O/A = 3.0	
	$N_{Di}/10^{-4}$	A in O, %	$N_{Di}/10^{-4}$	A in O, %	$N_{Di}/10^{-4}$	A in O, %
1.2 $\underline{M}$ TBP - Isopar™ - L	16.7	1.7	5.1	2.1	4.2***	2.2
1.2 $\underline{M}$ DA[AP] - Norpar™ - 12	13.2	0.4	5.1	1.0	4.6	2.0
1.2 $\underline{M}$ DA[AP] - Isopar™ - L	13.4	2.0	4.5	1.0	4.1	4.3
1.2 $\underline{M}$ DA[AP] - Isopar™ - M	7.4	9.0	2.7	3.0	2.6	4.0

* Dispersion tests were performed with 100 mL of solution except where noted.

** This percentage reflects the highest amount of aqueous phase entrained in the organic phase after phase disengagement. Usually three dispersion tests were performed per O/A per process solvent.

*** 60 mL of solution used to measure N_{Di}.

Table 8. Dispersion numbers*(N_{Di}) and aqueous entrainment (A in O)** for different process solvent formulations.
0.2 $\underline{M}$ CMPO - 0.2 $\underline{M}$ DtBuCH18C6 - Phase modifier-diluent Room temperature (23.5 to 25.0 °C)

Aqueous phase = 0.25 $\underline{M}$ Na$_2$CO$_3$

Phase Modifier-Diluent	O/A = 0.33		O/A = 1.0		O/A = 3.0	
	$N_{Di}/10^{-4}$	A in O, %	$N_{Di}/10^{-4}$	A in O, %	$N_{Di}/10^{-4}$	A in O, %
1.2 $\underline{M}$ TBP - Isopar™ - L	7.7	3.2	5.1	3.8	1.4***	25
1.2 $\underline{M}$ DA[AP] - Norpar™ - 12	7.1	4.0	6.5	4.0	4.2	4.0
1.2 $\underline{M}$ DA[AP] - Isopar™ - L	11.6	3.6	8.3	5.7	3.7	2.8
1.2 $\underline{M}$ DA[AP] - Isopar™ - M	8.1	5.0	6.3	5.0	1.5	10

* Dispersion tests were performed with 100 mL of solution except where noted.

** This percentage reflects the highest amount of aqueous phase entrained in the organic phase after phase disengagement. Usually three dispersion tests were performed per O/A per process solvent.

*** 60 mL of solution used to measure N_{Di}.

Combined Process Solvent: Chemical Properties

We showed earlier that the phase modifier (TBP or DA[AP]) and the chain length and branching of the hydrocarbon diluent have a major effect on the physical properties of the process solvent formulation. These same variables have only a minor effect on the distribution ratios of strontium, tectnetium, americium, and bismuth. Although the data presented in this section were not obtained with what we now regard as the optimum process solvent formulation, namely, 0.2 $\underline{M}$ DtBuCH18C6 - 0.2 $\underline{M}$ CMPO-1.2 $\underline{M}$ DA[AP] in Isopar™-L, the D values for all systems are very similar, regardless of whether TBP or DA[AP] is used as the phase modifier or whether Isopar™-L or M is used as the diluent. (For example, Table 9 compares the nitric acid dependency of D_{Am} using 0.2 $\underline{M}$ CMPO-1.2 $\underline{M}$ TBP-Isopar™-M and 0.2 $\underline{M}$ CMPO-1.2 $\underline{M}$ DA[AP]-Isopar™-M. Also compare D_{Sr}, D_{Am}, and D_{Tc} in Figures 4 and 5.)

Table 9. Influence of phase modifier on the acid dependency of D_{Am} Organic phase 0.2 $\underline{M}$ CMPO-1.2 $\underline{M}$ Phase modifier-Isopar™-M Temperature 24 °C

[HNO3], $\underline{M}$	D_{Am}	
	TBP	DA[AP]
0.05	0.26	0.24
0.10	0.71	0.84
0.25	4.8	4.4
0.51	11	8.5
1.0	20	20
2.4	43	33
3.1	36	29
4.0	32	26
6.0	26	17

Figure 4 shows the effect of two different temperatures, 25° and 50 °C, on the distribution ratios of strontium, americium, tectnetium, and bismuth vs. HNO3 concentration with the DtBuCH18C6-CMPO-TBP-Isopar™-L system. In every case, the distribution ratio is lower at the higher temperature. In the case of strontium and americium the distribution ratios are lowered by a factor of ~3. The distribution ratios of tectnetium and bismuth are less affected by temperature at the high nitric acid concentrations that are most likely to be present in the feed and where elevated temperature is most likely to occur.

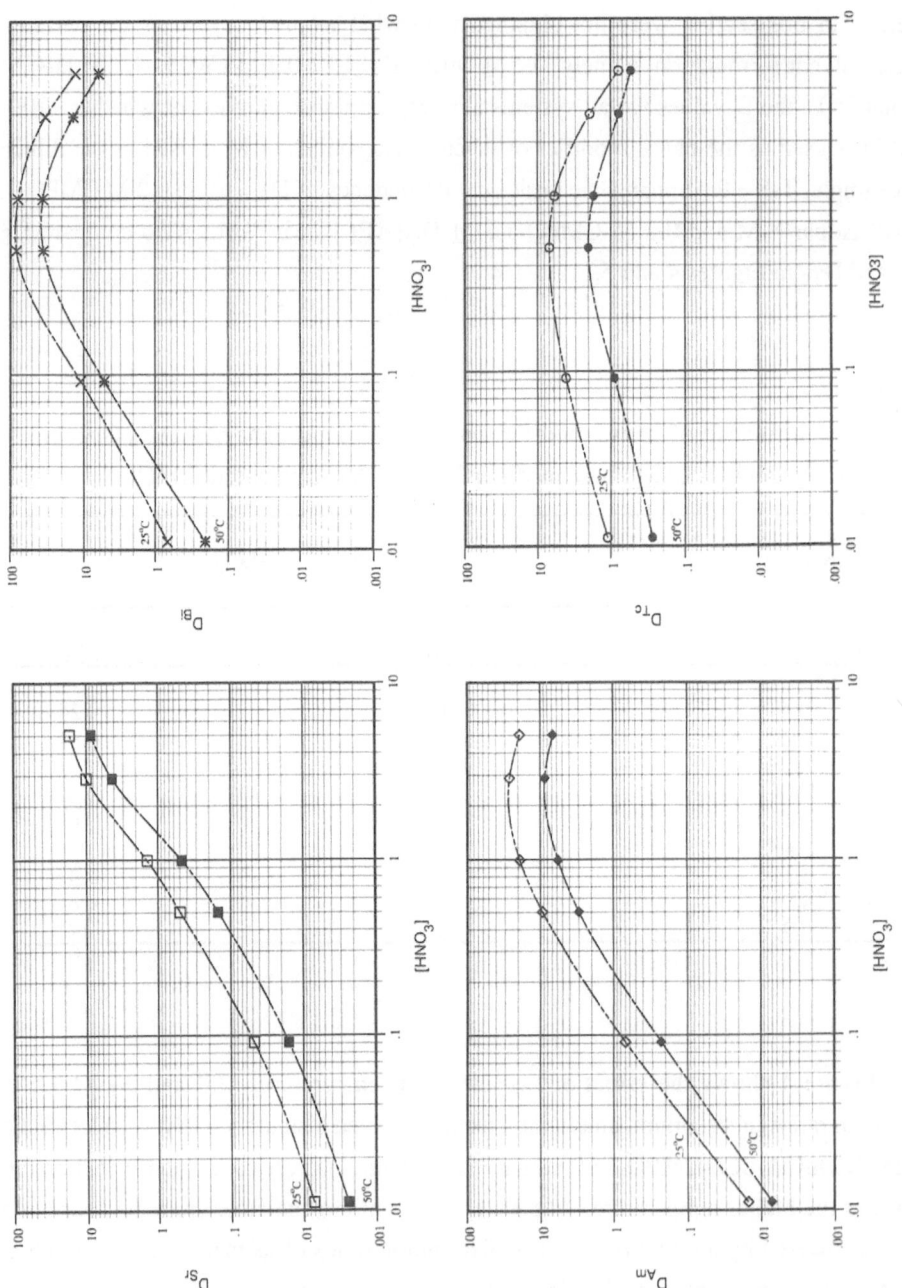

Figure 4. Acid dependency of D_{Am}, D_{Sr}, D_{Tc}, and D_{Bi} at 25 and 50° C for 0.2 $\underline{M}$ DBuCH18C6 -0.2 $\underline{M}$ CMPO - 1.2 $\underline{M}$ TBP in IsoparTM - L.

Figure 5 shows the influence of aqueous HNO₃ concentration on the extraction of strontium, tectnetium, and americium as measured by their distribution ratios using the DtBuCH18C6-CMPO-DA[AP]-Isopar™-M system. The measurements were made at room temperature (24 °C). Strontium is efficiently extracted above 2 $\underline{M}$ HNO₃ and americium, as is characteristic of TRUEX chemistry, is efficiently extracted over a wide range (0.5 to 6 $\underline{M}$) of HNO₃ concentrations. On the other hand, technetium is more effectively extracted at < 1 $\underline{M}$ HNO₃, but D_{Tc} is still high enough in 2 to 3 $\underline{M}$ HNO₃, which is the most likely concentration of acid in the dissolved sludge waste, to achieve the necessary decontaminations of tectnetium. (Most of the tectnetium is found in the alkaline supernatant in the waste tanks and not in the sludge.)

The influence of macro concentrations of uranium and bismuth on $D_{Sr, Tc,}$ americium is of equal importance to the data in Figure 5. Figure 6 shows that macro quantities of uranium and bismuth are very similar in their influence on the extraction of strontium, tectnetium, and americium. As expected, D_{Sr} is not significantly affected by either bismuth or uranium because neither is extracted to any degree by DtBuCH18C6 (Horwitz, et al., 1991). Americium extractability is decreased by macro concentrations of uranium and bismuth due to their extractability by CMPO. However, the diminution

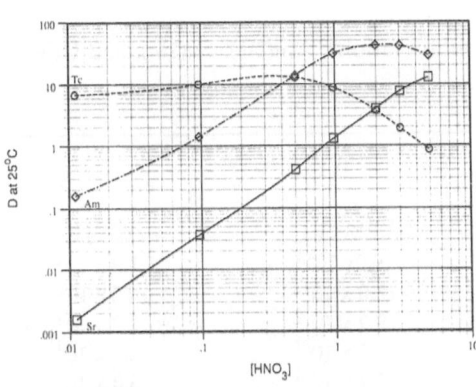

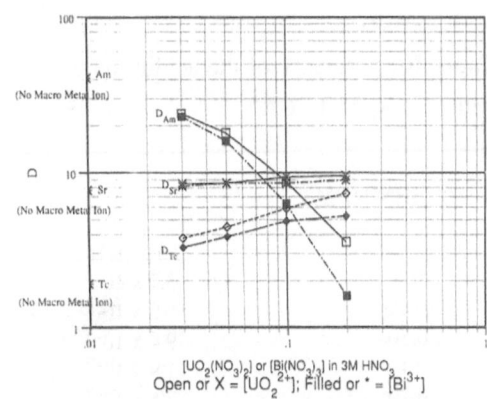

Figure 5. Acid dependency of D_{Sr}, D_{Sr}, and D_{Tc}, at 25 °C for 0.2 $\underline{M}$ DtBuCH18C6 -0.2 $\underline{M}$ CMPO - 1.2 $\underline{M}$ DA[AP] in Isopar™-M

Figure 6. The influence of loading on the extraction of Sr, Am, and Tc using 0.2 $\underline{M}$ DtBuCH18C6 -0.2 $\underline{M}$ CMPO - 1.2 $\underline{M}$ DA[AP] in Isopar™-M.

in D_{Am} with increasing concentration of uranium and bismuth is much less than expected based on UO₂(NO₃)•2CMPO and Bi(NO₃)₃•3CMPO stoichiometries. The explanation probably lies in the fact that DA[AP] replaces one of the CMPO molecules coordinated to uranium and bismuth, resulting in much less rapid consumption of CMPO. Technetium extraction is increased by increasing concentrations of uranium and bismuth due to complexation between TcO_4^- and UO_2^{2+} or Bi^{3+}.

The data in Figures 5 and 6 show that the DtBuCH18C6-CMPO-DA[AP] solvent combination in a branched chain hydrocarbon diluent is an effective system for extracting both strontium and tectnetium, as well as all of the actinides, from highly acidic nitrate media.

Selective Partitioning of Strontium and TRUs from Uranium and Tectnetium

An important feature of the combined TRU-Sr extraction process is the partitioning of strontium and TRUs from uranium and tectnetium. This selective partitioning can be readily achieved using a new stripping agent, tetrahydrofuran-2,3,4,5-tetracarboxylic acid, abbreviated THFTCA. Tables 10 and 11 show the D's for Am, Pu(IV), and U(VI) as a function of HNO_3 concentration in the presence of 0.25 and 0.50 $\underline{M}$ THFTCA, respectively. Data for strontium are not shown because D_{Sr} is sufficiently low below 0.2 $\underline{M}$ HNO_3 that no complexation by THFTCA is required. The data show a very effective partitioning of americium and plutonium from uranium. The higher of the two THFTCA concentrations is considered the better choice because of the lower values for D_{Am} and D_{Pu} with 0.5 $\underline{M}$ THFTCA. In addition, THFTCA has a small partition coefficient of ~0.6 between the combined process solvent and dilute nitric acid. Using a more concentrated THFTCA concentration ensures that the aqueous concentration is maintained at an effective level. After stripping strontium and TRUs, uranium and tectnetium are readily stripped from the combined process solvent using 0.25 $\underline{M}$ Na_2CO_3.

Table 10. Influence of the acidity on the distribution ratio of Am(III), Pu(IV), and U(VI) with 0.25 $\underline{M}$ THFTCA*

[HNO3], $\underline{M}$	D		
	Am(III)	Pu(IV)	U(VI)
0.0	3.7×10^{-3}	9.0×10^{-2}	1.9×10^1
0.025	6.9×10^{-3}	8.0×10^{-2}	1.7×10^1
0.50	9.4×10^{-3}	9.9×10^{-2}	2.0×10^1
0.10	1.9×10^{-2}	1.3×10^{-1}	2.7×10^1
0.20	7.5×10^{-2}	2.4×10^{-1}	3.0×10^1

*Organic phase: 0.2 $\underline{M}$ CMPO - 0.2 $\underline{M}$ DtBuCH18C6 - 1.2 $\underline{M}$ DA[AP] - Isopar^{TM}-M.

Table 11. Influence of the acidity on the distribution ratio of Am(III), Pu(IV), and U(VI) with 0.50 $\underline{M}$ THFTCA*

[HNO3], $\underline{M}$	D		
	Am(III)	Pu(IV)	U(VI)
0.0	3.0×10^{-3}	4.5×10^{-2}	8.9
0.025	4.2×10^{-3}	5.1×10^{-2}	9.2
0.050	6.5×10^{-3}	5.9×10^{-2}	9.5
0.10	1.2×10^{-2}	7.6×10^{-2}	1.2×10^1
0.20	4.7×10^{-2}	1.0×10^{-1}	2.0×10^1

*Organic phase: 0.2 $\underline{M}$ CMPO - 0.2 $\underline{M}$ DtBuCH18C6 - 1.2 $\underline{M}$ DA[AP] - Isopar^{TM}-M.

Flowsheet Development

A flowsheet (Figure 7) has been developed for testing of the combined TRUEX-SREX process. Initial tests of this flowsheet will focus on the nitric acid concentration at the various process stages in a 20-stage minicontactor. Using the Generic TRUEX Model (Vandegrift, et al., 1992) and, when appropriate, additional data specific to the combined TRUEX-SREX process, the flowsheet design will be evaluated to be sure that it meets process goals. The calculated concentration profiles for hydrogen, strontium, americium, plutonium, and tectnetium are given in Figures 8, 9, 10, 11 and 12, respectively.

The flowsheet will first be run with only nitric acid in the feed. The process solvent composition is 0.2 $\underline{M}$ DtBuCH18C6-0.2 $\underline{M}$ CMPO-1.2 $\underline{M}$ DA[AP] in Isopar™-L. Isopar™-L is used in place of -M because the former has higher dispersion numbers. Distribution ratios for the various solutions with the Isopar™-L formulation are identical, for all practical purposes, to those shown in Figures 5 and 6. Scrub 1 is 0.04 $\underline{M}$ HNO$_3$, the strip is 0.01 $\underline{M}$ HNO$_3$ and the carbonate wash is 0.25 $\underline{M}$ Na$_2$CO$_3$. After the contactor reaches steady state and HNO$_3$ concentration samples are collected from the Aqueous Raffinate and the Strip Effluents and from selected scrub and strip stages the Dissolved Sludge Waste and Scrub 1 feeds are changed to those shown in Figure 7. After steady state is reached, samples are collected to determine the effect of these other components on the HNO$_3$ concentration in the Strip Effluent. Finally, the Strip feed is changed to that shown in Figure 7. After steady state is reached, samples are collected to determine the THFTCA concentration in the Strip Effluent and the Carbonate Effluent. These tests are done to verify that (1) the HNO$_3$ concentration in the Strip Effluent is low enough for the THFTCA to effectively strip out americium, plutonium, and the lanthanides and (2) the loss of THFTCA to the Carbonate Wash is low.

Strontium is the key component in the extraction section. The model shows that the decontamination factor (D.F.) for strontium would be 1.2 x 10^4 for the six stages provided. Additional extraction stages for this flowsheet would increase the D.F. for strontium by a factor of 4.5 per stage. A plant design would use enough contactor stages in the extraction section to get the desired D.F. for strontium. Two or three more stages would then be added to accommodate process variations in concentration and flow rate.

One noteworthy feature of this flowsheet is the use of a split scrub section which has two functions. A concentrated Al(NO$_3$) solution removes oxalate and fluoride ions from the solvent and keeps them from reaching the strip section. In the second scrub solution, dilute HNO$_3$ removes most of the aluminum ions and much of the nitric acid. The key factor in the second scrub is the removal of nitric acid. The nitric acid concentration in the organic phase must be low enough that its concentration in the Strip Effluent is in the range of 0.2 to 0.3 $\underline{M}$ or lower. At this low HNO$_3$ concentration, the THFTCA concentration, which will be greater than 0.2 $\underline{M}$ in the aqueous phase for the first three strip stages, will be able to effectively strip americium, plutonium, and the lanthanides.

To recover plutonium in the strip section and to enhance the recovery of americium, plutonium, and the lanthanides, a 1 $\underline{M}$ THFTCA solution is used as the strip. Since its distribution coefficient is high, about 0.6, five strip scrub stages are used to keep the THFTCA in the strip section and to minimize the amount of THFTCA getting into the Carbonate Wash section. Using a distribution coefficient of 0.6 and the flowsheet shown in Figure 7, the THFTCA concentration in the process solvent leaving the strip scrub section would be only 0.024 $\underline{M}$. Additional strip scrub stages would reduce this THFTCA concentration even further.

Only one Carbonate Wash stage is shown in Figure 7. Two or three stages would be recommended for an actual plant design. In addition, an Acid Rinse section with a feed of 0.3 $\underline{M}$ HNO_3 should be used before the combined process solvent is recycled. This step insures that Na_2CO_3 does not enter the extraction section where it would partially neutralize the acid there.

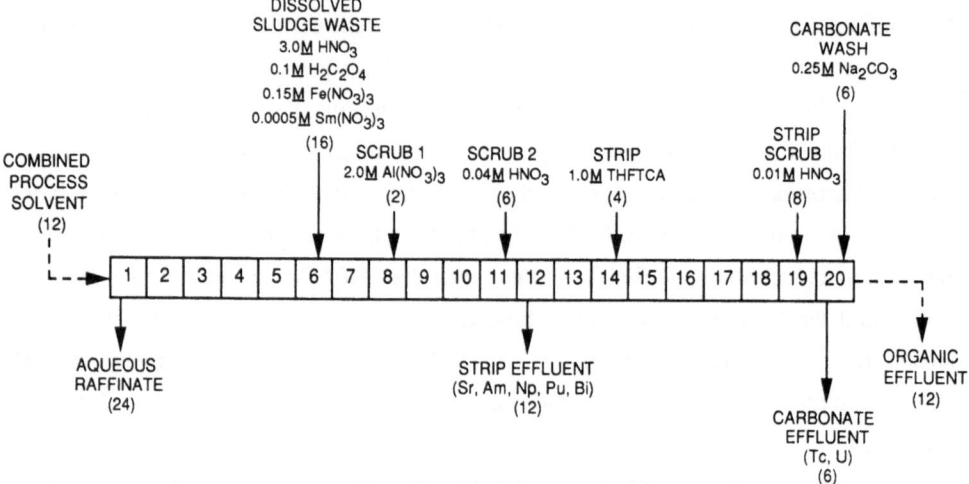

Figure 7. Flowsheet for testing the Combined TRUEX-SREX Process.

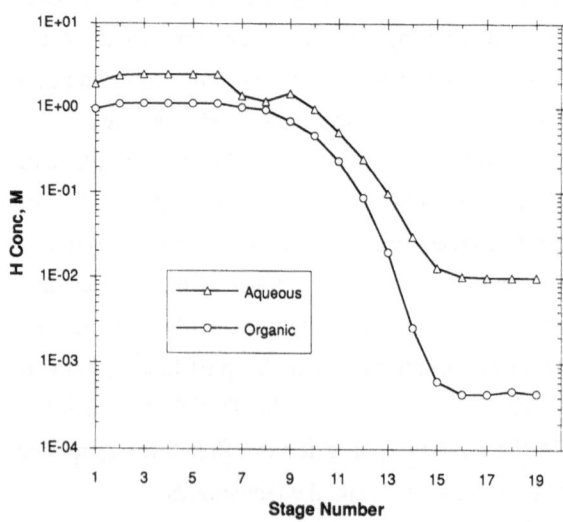

Figure 8. Calculated acid concentration for the various stages of the Combined TRUEX-SREX Process.

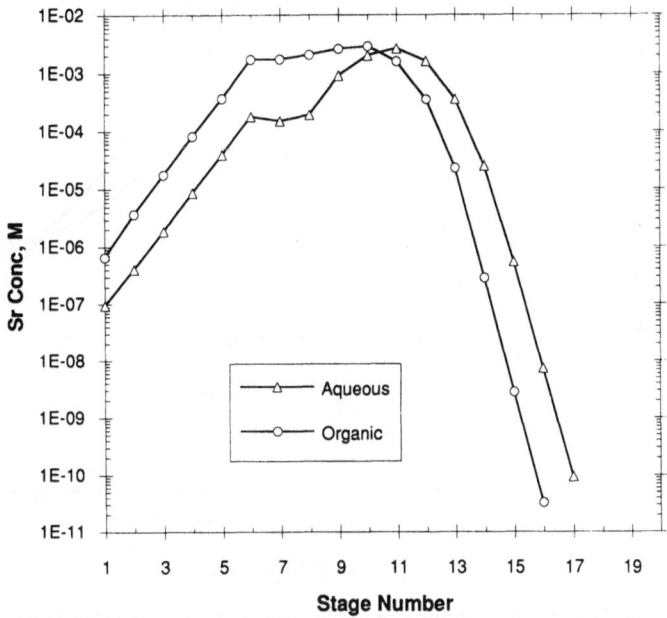

Figure 9. Calculated strontium concentration for the various stages of the Combined TRUEX-SREX Process.

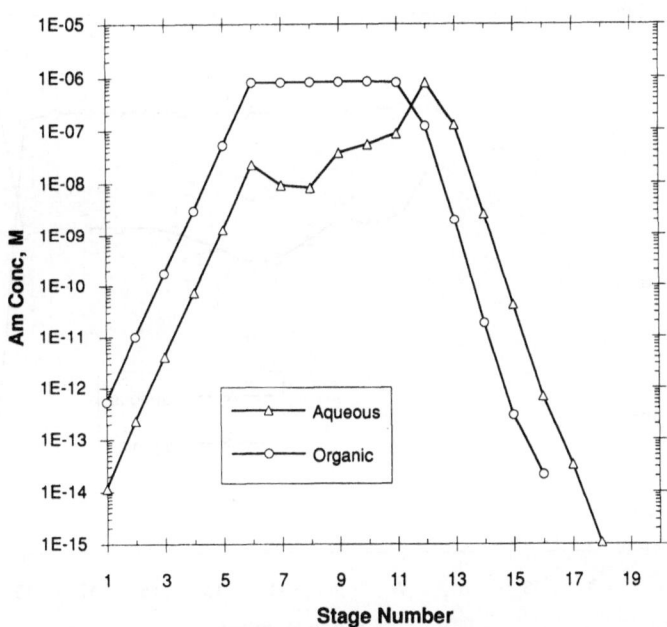

Figure 10. Calculated americium concentration for the various stages of the Combined TRUEX-SREX Process.

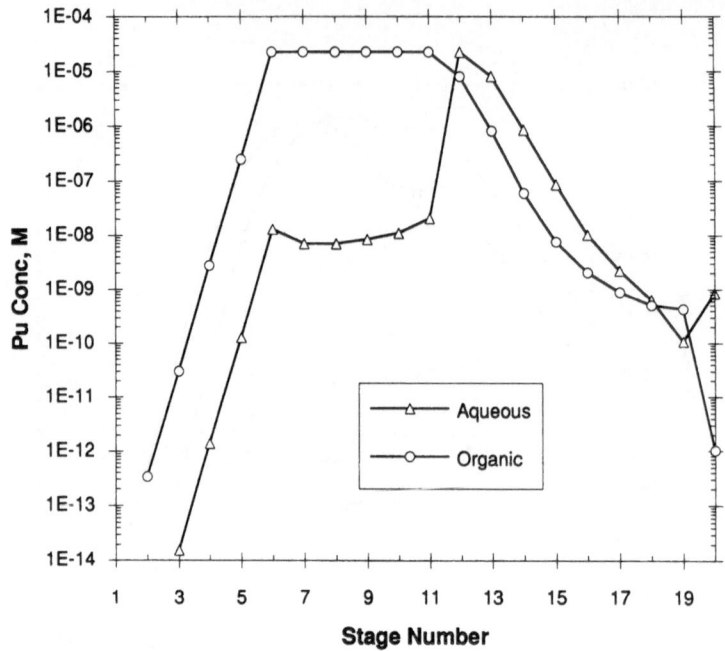

Figure 11. Calculated plutonium concentration for the various stages of the CombinedTRUEX-SREX Process.

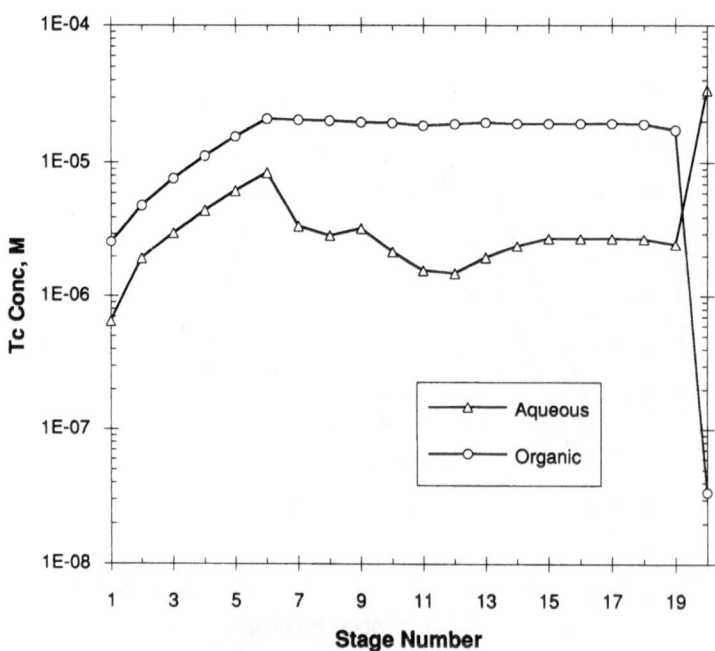

Figure 12. Calculated technetium concentration for the various stages of the Combined TRUEX-SREX Process.

ACKNOWLEDGEMENTS

The authors wish to thank the U. S. Department of Energy, Office of Technology Development (ESPIP) for financial support and the Westinghouse Hanford Co. for funding in the initial stages of this study that were performed in 1990-91. The authors also wish to thank Mr. Andrew Bond and Mr. Cary Bauer for assistance in the distribution ratio measurements and Drs. Claudia Fellinto and Hermi Brito for assistance with the third phase formation measurements.

REFERENCES

Chiarizia, R. and Horwitz, E.P., 1987, The influence of TBP on Am extraction by CMPO, *Inorg. Chim. Acta*, 140:261.

Gatrone, R.C., Kaplan,. L., and Horwitz, E.P., 1987, The synthesis and purification of the carbamoyl-methylphosphine oxides, *Solvent Extr. Ion Exch.*, 5:1075.

Horwitz, E.P., Kalina, D.G., Kaplan, L., Mason, G.W., and Diamond, H., 1982, Selected alkyl(phenyl)-N,N-dialkylcarbamoylmethylphosphine oxides as extractants for Am(III) from nitric acid media, *Separation Sci. Tech.*, 17:1261.

Horwitz, E.P. and Schulz, W.W., 1991, The TRUEX Process: A Vital Tool for Disposal of U. S. Defense Nuclear Waste, *in* "New Separation Chemistry Techniques for Radioactive Waste and Other Specific Applications," L. Cecille, M. Casarci and L. Pietrelli, eds., Elsevier Applied Science, London.

Horwitz, E.P., Dietz, M.L., and Fisher, D.E., 1991, SREX : A new process for the extraction and recovery of Sr from acidic nuclear waste streams, *Solvent Extr. Ion Exch.*, 9:1.

Horwitz, E.P., 1992, Combining Extractant Systems for the Simultaneous Extraction of Transuranic Elements and Selected Fission Products, *in* "Proceedings of the First Hanford Separation Science Workshop, Richland," Washington, July 23-25, 1991, PNL-SA-21775, May 1993, p. II.29.

Kolarik, Z. and Horwitz, E.P., 1988, Extraction of metal nitrates with octyl(phenyl)-N,N-diisobutylcar-bamoylmethylphosphine oxide in alkaline diluents at high solvent loading, *Solvent Extr. Ion Exch.*, 6:61.

Leonard, R.A., Bernstein, G.W., Pelto, R.H., and Ziegler, A.A., 1982, Liquid-liquid dispersion in turbulent couette flow, *AIChE Journal* 27: 495.

Vandegrift, G.F., 1992, Chapter II, Separation Science and Technology, *in* "Nuclear Technology Programs, Semiannual Progress Report," U. S. DOE Report ANL-91/42, 1992, 54, October 1989-March 1990, Argonne National Laboratory, Argonne, Illinois.

NOBLE METAL FISSION PRODUCTS AS CATALYSTS FOR HYDROGEN EVOLUTION FROM FORMIC ACID USED IN NUCLEAR WASTE TREATMENT

R.B. King,[1] A.D. King, Jr.,[1] N.K. Bhattacharyya,[1]
C.M. King,[2] and L.F. Landon[2]

[1] Department of Chemistry
University of Georgia
Athens, Georgia 30602
and
[2] Westinghouse Savannah River Technology Center
Aiken, South Carolina 29808

INTRODUCTION

One promising method for the disposal of highly radioactive defense nuclear wastes is a vitrification process in which the wastes are incorporated into borosilicate glass logs, the logs are sealed into welded stainless steel canisters, and the canisters are buried in suitably protected burial sites. The key features of this general scheme are outlined in Figure 1.

Operation of a glass melter and durability of the glass is affected by the redox state of the glass during processing. Formation of a conductive metallic sludge in an overreduced melt can result in a shortened melter lifetime. An overoxidized melt may lead to foaming and loss of ruthenium as volatile RuO_4; historically, foaming in the melter has been controlled by introduction of a reductant into the melter feed. For these reasons formic acid was selected by the Savannah River Laboratory (SRL) for its Defense Waste Processing Facility (DWPF) feeds as an acid which not only solubilizes the metals but also reduces Hg^{2+} and Hg_2^{2+} to mercury metal for subsequent steam stripping and manganese. Formic acid also reduces manganese in MnO_2 to Mn^{2+}; manganese is believed to contribute to reboil in the melt. Formic acid also decreases the melter feed viscosity thereby reducing the pumping needs for the plant.

Chemical Pretreatment of Nuclear Waste for Disposal, Edited by
W.W. Schulz and E.P. Horwitz, Plenum Press, New York, 1995

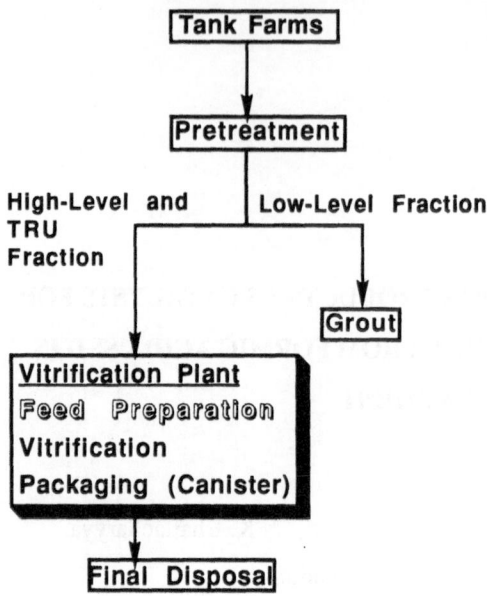

Figure 1. Flow chart for the vitrification of nuclear wastes.

For these reasons laboratory and pilot scale studies have been carried out both at SRL and the Battelle Pacific Northwest Laboratory (PNL) using formic acid for the pretreatment of simulated nuclear wastes. However, in 1988 hydrogen generation was observed at PNL during treatment of simulated high-level waste with formic acid in a laboratory-scale apparatus.[1] In pilot-scale studies at SRL in 1990 significant levels (three to four times the lower flammability limit) of hydrogen were observed during preparation of a DWPF feed simulant.

The suggested cause of this unwanted hydrogen generation is the catalytic decomposition of formic acid, i.e.,

$$HCO_2H \longrightarrow H_2\uparrow + CO_2\uparrow \tag{1}$$

This reaction does not proceed at a significant rate under the conditions encountered during the pretreatment of nuclear wastes in the absence of catalysts. However, possible catalysts for this reaction include the noble metals ruthenium, rhodium, and palladium in uranium fission products present, respectively, at levels of ~2.1, ~0.4, and ~1.2 kilograms per ton of spent uranium fuel, originally enriched to 3.3% in ^{235}U.[2] This paper describes laboratory-scale studies at the University of Georgia whose objective was to establish the role of noble metal fission products in catalyzing the decomposition of formic acid to hydrogen in nuclear waste media and finding methods of inhibiting this unwanted hydrogen generation. All of the experiments at the University of Georgia were performed using nonradioactive nuclear waste simulants provided by SRL.

There is a strong precedent in the literature for the noble metal catalyzed decomposition of formic acid to hydrogen. The earliest studies by Müller and Loerpabel[3] used salts of the six platinum group metals as catalyst precursors to effect the dehydrogenation of 10% formic acid containing dissolved sodium formate. More recent studies have focussed on various aspects of palladium-catalyzed formic acid decomposition. These studies include work by Ruthven and Upadhye[4] on the catalytic decomposition of aqueous formic acid over palladium black, work by Aguiló[5] on the palladium(II) catalyzed oxidation of formic acid in acetic acid solution, and work by Hill and Winterbottom[6] on the palladium-catalyzed decomposition of formic acid/sodium formate solutions. These catalyst systems all appear to be heterogeneous. In addition, soluble derivatives of other noble metals including ruthenium carbonyls[7] and the rhodium phosphine complex $Rh(C_6H_4PPh_2)$-$(PPh_3)_2$ (ref. 8) have been shown to be active homogeneous catalysts for formic acid decomposition. However, these soluble ruthenium and rhodium derivatives contain carbonyl or phosphine ligands which are expected to modify greatly the underlying noble metal chemistry and which are not present in the nuclear wastes being treated.

EXPERIMENTAL SECTION

Hydrogen and nitric oxide analyses were performed using a Varian 90P gas chromatograph. These gases were eluted on a 2.4 m × 6 mm column packed with 40/60 mesh 13× molecular sieves obtained from Varian Corporation. Argon was used as the carrier gas and the column temperature was maintained at 80°C. Carbon dioxide, nitric oxide, and nitrous oxide were monitored using a Fisher Model 1200 gas partitioner. The gases were separated on the basis of their size and polarity by means of two columns, a 2 m 80/100 mesh Columpak PQ and a 3.3 m 13× molecular sieve column mounted in series. Helium was used as the carrier gas. The temperatures of both columns were maintained at 50°C. Data processing was performed using a Hewlett-Packard HP 3396 Series II integrator.

The sludge simulants used for this study were obtained from the SRL stock and were prepared by an outside contractor. The compositions of the sludge simulants are given in Table 1. Noble metal compounds were obtained from standard commercial suppliers.

In a typical experiment 38.5 mL of the sludge simulant and 1.5 mL of 88% formic acid were placed in a 250 mL glass reaction vessel equipped with a rubber septum for gas sampling. The reaction mixture was then flushed with argon. A weighed quantity of the noble metal compound (e.g., $RhCl_3 \cdot 3H_2O$) was then added to the reaction vessel against a countercurrent flow of argon. The reaction mixture was stirred at room temperature with a magnetic stirrer. The gas phase was sampled and analyzed by gas chromatography. Subsequently the reaction mixture was heated to the desired temperature (80–100°C) and gas samples were analyzed periodically by gas chromatography. The pressure of the system was monitored by a pressure gauge attached to the reactor. Pressures greater than 3 atm were avoided owing to leakage problems.

These experiments were performed in closed (static) systems rather than the open (flow) systems to be used in the actual processing and currently used in pilot studies at SRL and PNL. Closed systems were chosen for this study in order to retain all gaseous components in the system to allow a more accurate material balance for the determination of stoichiometries. However, gaseous products (e.g., NO, and N_2O) formed in early stages of reactions remain in the closed system and thus are available for later reactions. The time derivatives $(\partial f/\partial t)$ of the gas evolution (f) versus time (t) curves obtained in the closed systems used in this work correspond to the gas evolution versus time curves obtained in

Table 1. Compositions of the IDMS and PUREX process sludge simulants used in this work

	IDMS HM Sludge	Low Nitrite PUREX Process Sludge
	weight %	weight %
$Al(OH)_3$	25.031	12.298
$BaSO_4$	0.219	0.505
$Ca_3(PO_4)_2$	0.097	0.254
$CaCO_3$	1.478	5.148
$CaSO_4$	0.000	0.468
Cr_2O_3	0.220	0.409
$CsNO_3$	0.028	0.003
CuO	0.054	0.182
$Fe(OH)_3$	25.893	51.204
Nd_2O_3	2.342	0.253
KOH	0.170	0.397
MgO	0.342	0.246
MnO_2	6.650	7.320
Na_2CO_3	0.107	0.148
Na_2SO_4	0.471	0.169
Na_3PO_4	0.041	0.012
$NaCl$	0.476	1.545
NaF	0.259	0.275
NaI	0.000	0.027
$NaNO_2$	12.083	2.455
$NaNO_3$	0.478	0.358
$NaOH$	1.367	3.507
$Ni(OH)_2$	1.308	4.507
$PbSO_4$	0.220	0.505
SeO_2	0.002	0.003
SiO_2	5.039	1.672
$SrCO_3$	0.408	0.038
TeO_2	0.028	0.031
Zeolites	9.342	1.661
ZnO	0.043	0.341
ZrO_2	1.842	3.828

open flow systems. Thus, a constant H_2 concentration measured in a closed system corresponds to a peak decaying to zero in a corresponding open flow system.

RESULTS

The sludge simulants used in this work (Table 1) have the following features that are relevant to their reactivity towards formic acid:

(1)   The presence of considerable amounts of carbonate and nitrite leads to copious gas evolution (CO_2, NO, N_2O) upon treatment with formic acid according to the equations:

$$2\ HCO_2H + CO_3^{2-} \longrightarrow \boxed{CO_2} + 2\ HCO_2^- + H_2O \tag{2}$$

$$HCO_2H + NO_2^- \rightleftharpoons \{HNO_2\} + HCO_2^- \tag{3a}$$

$$3HNO_2 \rightleftharpoons 2\ \boxed{NO} + HNO_3 + H_2O \tag{3b}$$

$$2\ HCO_2H + 2\ HNO_2 \longrightarrow \boxed{N_2O} + 2\ \boxed{CO_2} + 3\ H_2O \tag{4}$$

Reaction 3b involves disproportionation of nitrous acid generated from nitrite and formic acid whereas reaction 4 involves reduction of nitrous acid with formate ion. Figure 2a shows that both of these reactions occur upon treatment of pure nitrite with formic acid as indicated by the production of CO_2, N_2O and NO whereas figure 2b shows that treatment of pure nitrite with the non-reducing acetic acid leads as expected only to disproportionation as indicated by the production of only NO.[9]

(2) The presence of redox-active components, particularly iron(III), in the sludge simulant can affect the relevant noble metal chemistry.

(3) The presence of nitrite in the sludge simulant can also affect the noble metal chemistry by forming soluble metal nitrosyls or nitro complexes (Figure 3), which can affect the catalytic activities and/or reduction potentials of the noble metals.

In the initial phase of this work, the three noble metals of interest (Ru, Rh, Pd) were screened for catalytic activity using the IDMS HM sludge simulant (Table 1). The catalytic activities of these three noble metals, introduced as their chlorides, for the decomposition of formic acid to H_2 under various conditions are summarized in Table 2. These results indicate that the catalytic activities of the noble metals for decomposition of formic acid in the sludge simulant media are very different than those observed with pure formic acid. Thus in the sludge simulant media rhodium is a very active catalyst for formic acid decomposition whereas rhodium is not an active catalyst for the decomposition of pure formic acid under similar conditions. The results in Table 2 indicate that nitrite ion found in the sludge simulant media functions as a promoter for the catalytic activity of rhodium.

Figure 4 shows the evolution of H_2 and CO_2 as a function of time for treatment of the low nitrite (~2.5% $NaNO_2$—see Table 1) PUREX process sludge simulant with formic acid in the presence of a rhodium catalyst. Hydrogen evolution occurs after an induction period (~200 min in this experiment). The expected copious CO_2 evolution from the carbonate sludge simulant as well as from formic acid decomposition (equation 1) also occurs.

Addition of sodium nitrite to the low-nitrite sludge simulant (Figure 4) produces both NO and N_2O (Figure 5). In accordance with equations 3b and 4 NO produced initially is consumed upon further reaction.

The effect of increasing the nitrite concentration on the rhodium-catalyzed hydrogen generation rate from formic acid in sludge simulant media is summarized in Figure 6. Note that relatively little H_2 (~0.3 mmoles at the end of the experiment) is produced in a PUREX process sludge simulant containing all of the components listed in Table 1 except for the nitrite. The H_2 generation behavior from 38.5 mL of nitrite-free sludge simulant with 625 mg of added sodium nitrite is essentially equivalent to that of PUREX process sludge containing the usual amount of nitrite (Table 1).

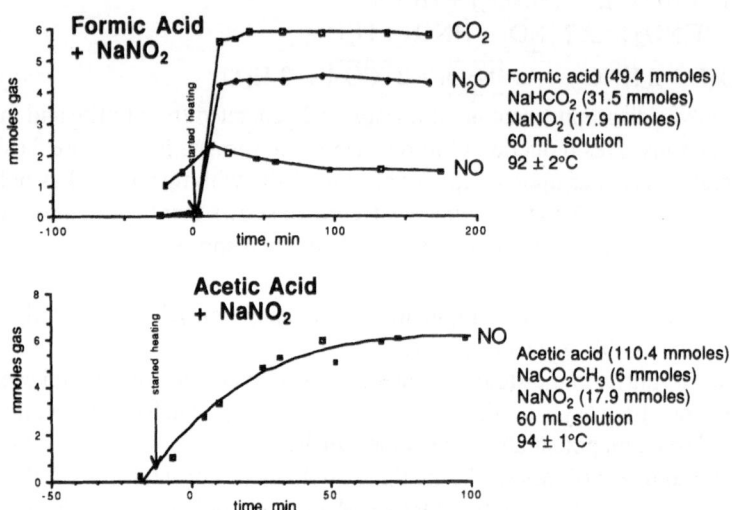

Figure 2. Gases evolved upon treatment of sodium nitrite with formic acid and with acetic acid.

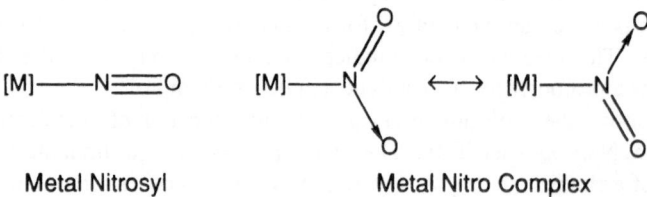

Metal Nitrosyl   Metal Nitro Complex

Figure 3. General structures of metal nitrosyls and metal nitro complexes.

Table 2. Summary of the catalytic activity of noble metals in the decomposition of formic acid to hydrogen at $90 \pm 5°C$*

Metal Source →	Ruthenium Hydrated RuCl$_3$ (0.22±0.01 mmoles)	Rhodium Hydrated RhCl$_3$ (0.21±0.01 mmoles)	Palladium PdCl$_2$ (0.45±0.02 mmoles)
Medium ↓			
3.3% aq.formic acid	4	0	3
0.76 M aq. NaCO$_2$H	40	2.4	78
3% HCO$_2$H/NaNO$_2$	3	118	5
HM sludge simulant	0	112	8

*Numbers are moles of H$_2$ produced per 24 hr day per mole of noble metal.

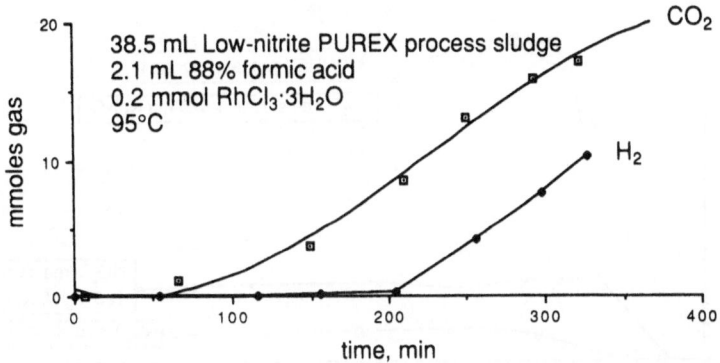

Figure 4. Hydrogen evolution from formic acid in low nitrite (~2.5% NaNO$_2$) PUREX process sludge simulant using a rhodium catalyst.

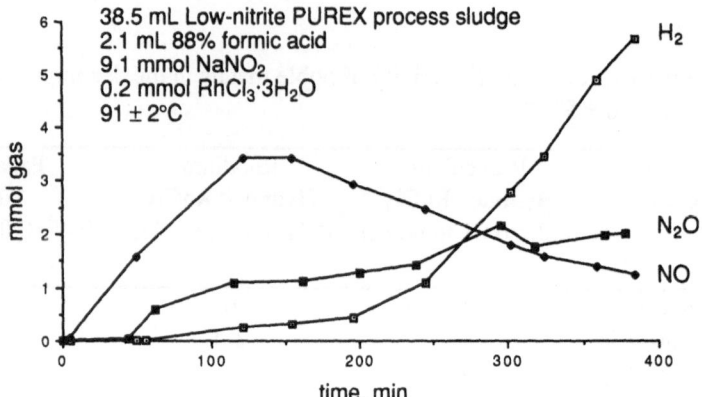

Figure 5. Effect of added nitrite on the rhodium-catalyzed hydrogen evolution from low nitrite PUREX process sludge simulant. The curve depicting the production of CO_2 is omitted for clarity.

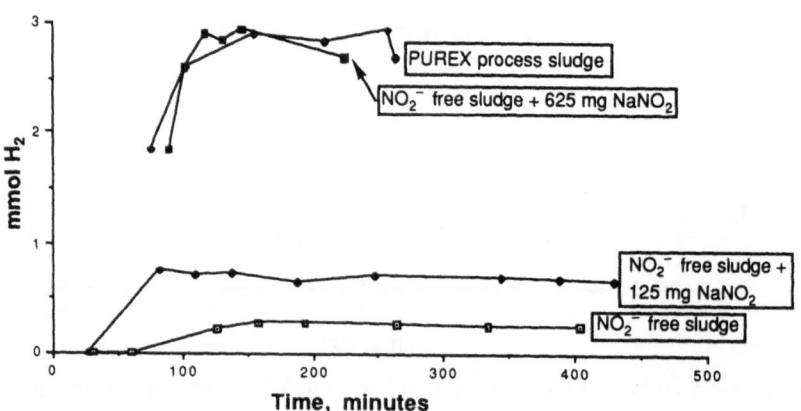

Figure 6. The effect of nitrite ion on the rhodium-catalyzed H_2 evolution from formic acid in sludge simulants.

Some experiments were performed using a nitrite-free sludge simulant and other noble metals of interest. Ruthenium, introduced as commercial "hydrated ruthenium trichloride," catalyzes the evolution of H_2 from formic acid in the nitrite-free sludge simulant but only after a much longer induction period (~1200 min) than observed in any of the rhodium experiments (Figure 7). Thus nitrite ion, although a promoter for the rhodium-catalyzed evolution of hydrogen from formic acid, is an inhibitor for ruthenium-catalyzed evolution of hydrogen from formic acid.

The considerable catalytic activity of rhodium for H_2 evolution from formic acid in sludge simulant media led to a search for inhibitors for this reaction. Table 3 shows that the aminopolycarboxylic acids (Figure 8) EDTA (ethylenediamine tetraacetic acid) and NTA (nitrilotriacetic acid) can inhibit the rhodium-catalyzed H_2 generation from formic acid. NTA is more effective in lower amounts than EDTA. The results summarized in Table 3 suggest that the use of formic acid containing ~10% NTA for the processing of nuclear wastes should result in considerably less H_2 evolution than the use of pure formic acid.

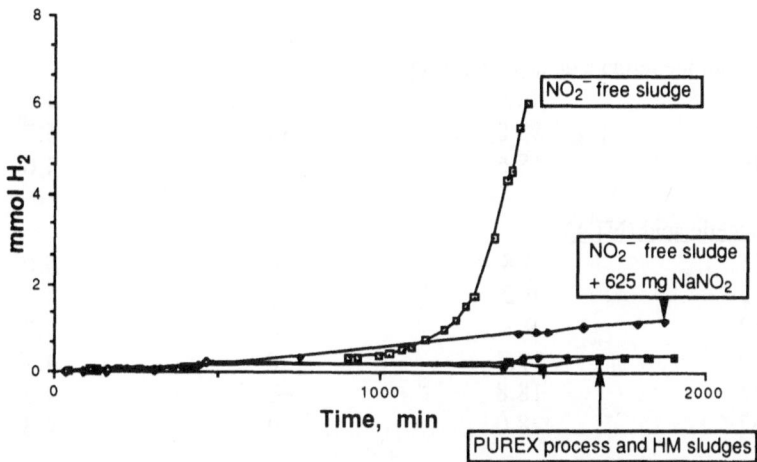

Figure 7. Effect of nitrite on the ruthenium-catalyzed hydrogen evolution from formic acid in sludge simulants.

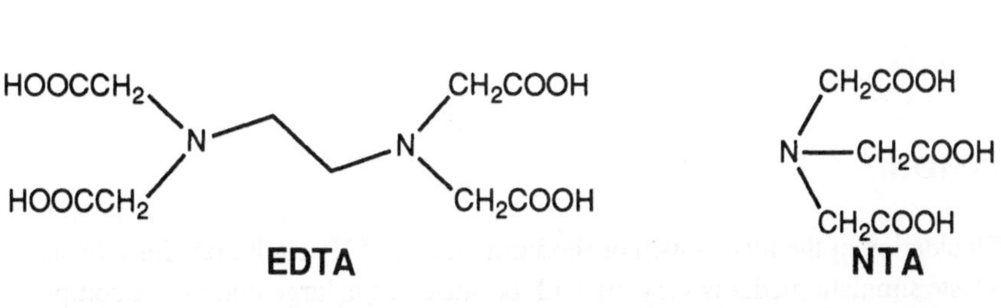

Figure 8. Chemical structures of the aminopolycarboxylic acids EDTA and NTA.

Several complexing agents of types other than aminopolycarboxylic acids were also investigated as inhibitors for the rhodium-catalyzed H_2 evolution from formic acid in sludge simulant media (Table 4). The complexing agents dimethylglyoxime, oxalic acid, sodium sulfide, and trimethyl phosphite are all ineffective as H_2 generation inhibitors. The only effective H_2 generation inhibitor in addition to the aminopolycarboxylic acids EDTA and NTA (Table 3) found in these studies was mercury. Mercury could be introduced as $HgCl_2$, as was done in the experiments summarized in Table 4, or even as elemental mercury. These small scale experiments suggest that significant rhodium-catalyzed H_2 evolution from the formic acid processing of nuclear wastes should only occur after the mercury is removed by reduction followed by steam distillation.

Table 3. Inhibition of hydrogen production with aminopolycarboxylic acids*

mmoles additive	mmoles additive/Rh	H_2 turnover / day†	pH at end of expt.
A. No additive			
0	0	80	6.2
B. Ethylenediamine tetraacetic acid (EDTA)			
0.52	6.7	85	4.8
0.77	13.8	35	5.7
1.04	18.6	~0	4.0
C. Nitrilotriacetic acid (NTA)			
0.21	3.8	90	5.0
0.35	6.2	~0	4.7
0.52	9.3	~0	4.7
0.99	17.7	~0	4.7
1.05	18.8	~0	4.3
1.57	28.0	~0	4.3

*All experiments were done with 30 mL of low nitrite PUREX process sludge simulant and 1.6 mL of 88% formic acid with 0.056 mmoles of $RhCl_3 \cdot H_2O$ at $100\pm2°C$.
†The designation "~0" means that negligible H_2 was produced after waiting at least 5 hr.

DISCUSSION

Elucidation of the mechanism of rhodium-catalyzed H_2 evolution from formic acid in nuclear waste simulant media is very difficult because of the large number of components in the nuclear waste simulants and the limited catalyst lifetime. Both homogeneous and heterogeneous mechanisms can be considered for this reaction.

A homogeneous mechanism for the rhodium-catalyzed H_2 evolution from formic acid involves an unstable soluble rhodium complex as the catalytically active species. The limited catalyst lifetime coupled with the complexity of the sludge simulants has prevented us from attempting to isolate a catalytically active rhodium species. The role of nitrite ion as a promoter for this catalytic reaction suggests the involvement of nitrorhodium complexes. Such nitrorhodium complexes can be obtained by substitution of up to three nitro groups in the hexanitrorhodate ion, $Rh(NO_2)_6^{3-}$, initially formed from nitrite ion and hydrated rhodium trichloride, with other ligands, including formate ion. This chemistry can lead to the mechanism summarized in Figure 9. The feasibility of such a mechanism is suggested by the observation more than a century ago[10] that rhodium trichloride reacts with excess nitrite ion

Table 4. Evaluation of inhibitors for the Rh-catalyzed generation of hydrogen from formic acid.

mmoles Rh as $RhCl_3 \cdot 3H_2O$	mmoles added $NaNO_2$	Additive	mmoles additive	Temp °C	H_2 Turnover per 24 hr.*
A. Experiments with 0.76 M Aqueous Formic Acid					
0.22	1.16	none	0	90	118
0.21	1.18	dimethylglyoxime	0.44	90	167
0.19	1.17	oxalic acid	0.41	95	79
0.20	1.17	EDTA	0.43	85	0
B. Experiments with the IDMS Sludge Simulant†					
0.18	0	none	0	82	112
0.19	0	Na_2S	2.17		high
0.21	0	TeO_2	2.0	80	62
0.20	0	$(MeO)_3P$	2.0	82	117
0.13	0	$HgCl_2$	0.97	85	7
0.2	0	$HgCl_2$	2.0	96	1.3
0.2	0	$HgCl_2$	19.6	96	0.01
0.2	0	EDTA	21.5	95	0
0.2	0	EDTA	2.2	96	16

* mmoles H_2/mmoles Rh/24 hr.
†This feed simulant is nominally ~0.3 M in nitrite (Table 1).

in boiling water to give salts of the hexanitrorhodate ion, $[Rh(NO_2)_6]^{3-}$ and the subsequent observation that three of the six nitro groups in the hexanitrorhodate ion are reported to be labile upon treatment with strong oxidants[11] or amidosulfuric (sulfamic) acid.[12]

A heterogeneous mechanism for the rhodium-catalyzed H_2 evolution from formic acid can involve small catalytically active rhodium metal clusters obtained by reduction of soluble rhodium compounds to rhodium metal as the catalytically active species. Such a heterogeneous catalyst can be deactivated by aggregation to larger particles. Non-specific effects of added ions in affecting aggregation of the small rhodium metal clusters are anticipated. Heterogeneous catalysis of this type can be prevented by inhibiting the reduction of soluble rhodium compounds to the metal.

$$RhCl_3(H_2O)_3 + 6\ NO_2^- \longrightarrow [Rh(NO_2)_6]^{3-} + 3\ H_2O + 3\ Cl^- \qquad (1)$$

Reported preparative method[10] for $M_3[Rh(NO_2)_6]$ salts
(M = Na, K, NH_4, etc.)

Substitution equilibria
$$[Rh(NO_2)_6]^{3-} + H_2O \rightleftharpoons [Rh(NO_2)_5(H_2O)]^{2-} + NO_2^- \qquad (2)$$
$$[Rh(NO_2)_5(H_2O)]^{2-} + H_2O \rightleftharpoons [Rh(NO_2)_4(H_2O)_2]^- + NO_2^- \qquad (3)$$
$$[Rh(NO_2)_4(H_2O)_2]^- + H_2O \rightleftharpoons Rh(NO_2)_3(H_2O)_3 + NO_2^- \qquad (4)$$

Introduction of Formate
$$[Rh(NO_2)_4(H_2O)_2]^- + HCO_2^- \rightleftharpoons [Rh(NO_2)_4(H_2O)(HCO_2)]^{2-} + H_2O \qquad (5)$$

Decomposition of Formate
$$[Rh(NO_2)_4(H_2O)(HCO_2)]^{2-} \longrightarrow [Rh(NO_2)_4(H_2O)H]^{2-} + CO_2 \qquad (6)$$
$$[Rh(NO_2)_4(H_2O)H]^{2-} + H_2O \longrightarrow [Rh(NO_2)_4(H_2O)(OH)]^{2-} + H_2 \qquad (7)$$
$$[Rh(NO_2)_4(H_2O)(OH)]^{2-} + HCO_2H \longrightarrow [Rh(NO_2)_4(H_2O)_2]^- + HCO_2^- \qquad (8)$$

Sum of Equations 5-8
$$HCO_2H \longrightarrow CO_2 + H_2 \qquad (9)$$

Figure 9. Possible role of nitrorhodium complexes in the homogeneous catalysis of hydrogen production from formic acid.

The experimental evidence favoring homogeneous over heterogeneous catalysis or vice versa for the rhodium-catalyzed H_2 evolution from formic acid is ambiguous and can best be interpreted as suggestive of a nitrorhodium catalyst immobilized onto the substantial amounts of hydrous iron and aluminum oxides present in the sludge simulants. Replicate experiments with the PUREX process sludge simulant exhibit sufficient rate variations and an ill-defined dependence on the amount of rhodium introduced indicative of a solid phase catalyst rather than a homogeneous catalyst in solution. However, the specific promoter effect of nitrite ion suggests nitrorhodium complexes as catalytically active species rather than small rhodium metal clusters.

ACKNOWLEDGMENT

We are indebted to the U. S. Department of Energy for support of this work at the University of Georgia through the Westinghouse Savannah River Technology Center, the South Carolina Universities Educational and Research Foundation, and the University of South Carolina in Columbia. We would like to acknowledge the help of Ms. G. Vemparala in obtaining the data depicted in Figure 2.

REFERENCES

1. K. D. Wiemers, *Evaluation of Process Off-Gases Released During the Formating of Simulated HWVP Feed*, presented at the American Institute of Chemical Engineers 1988 Summer National Meeting, Denver, Colorado, August, 1988.
2. G. R. Choppin and J. Rydberg, *Nuclear Chemistry: Theory and Applications*, Pergamon, Oxford, 1980, p 505.

3. E. Müller and W. Loerpabel, Die katalytische Zersetzung wässeriger Lösungen von Ameisensäure durch die Platinmetalle IV, *Monatshefte*, 53/54: 825 (1929).
4. D. M. Ruthven and R. S. Upadhye, The catalytic decomposition of aqueous formic acid over suspended palladium catalysts, *J. Catal.*, 21: 39 (1971).
5. A. Aguilò, Pd(II)-catalysed oxidation of formic acid in acetic acid solution, *J. Catal.*, 13: 283 (1969).
6. S. P. Hill and J. M. Winterbottom, The conversion of polysaccharides to hydrogen gas. Part I: The palladium catalysed decomposition of formic acid/sodium formate solutions, *J. Chem. Tech. Biotechnol.*, 41: 121 (1988).
7. R. M. Laine, R. G. Rinker, and P. C. Ford, Homogeneous catalysis by ruthenium carbonyl in alkaline solution: the water gas shift reaction, *J. Am. Chem. Soc.*, 99: 252 (1977).
8. S. H. Strauss, K.H. Whitmire, and D. F. Shriver, Rhodium(I) catalyzed decomposition of formic acid, *J. Organometal. Chem.*, 174: C59 (1979).
9. G. Vemparala, M. S. Thesis, University of Georgia, 1992.
10. E. Leidié, Recherches sur les nitrites doubles du rhodium, *Compt. Rend.*, 111: 106 (1890).
11. A. V. Belyaev, Oxidation of sodium hexanitrorhodate(III) by strong oxidants, *Russ. J. Inorg. Chem.*, 12: 577 (1967).
12. Yu. N. Kukushkin and O. V. Stefanova, Kinetics of the reaction of platinum, palladium, rhodium, and cobalt nitrocomplexes with amidosulphuric acid, *Russ. J. Inorg. Chem.*, 22: 1844 (1977).

MICROBIOLOGICAL TREATMENT OF RADIOACTIVE WASTES

A. J. Francis

Department of Applied Science
Brookhaven National Laboratory
Upton, NY 11973

INTRODUCTION

A major national concern is the remediation of contaminated materials, soils, and water as well as disposal of wastes containing radionuclides and toxic metals safely and economically. Large volumes of wastes containing toxic metals and radionuclides are generated by nuclear- and/or fossil-fueled power plants, the metal fabrication industries, and by facilities producing nuclear weapons. New innovative treatment and remediation technologies are needed because the problem is pervasive. Options for microbiological treatment include stabilization, or the removal and recovery of the contaminants. Stabilization means that the radionuclides and toxic metals are converted chemically or biologically to an insoluble form which is stable in the environment. For decontamination, both metal and radionuclide must be removed and recovered from the contaminated site, so that the site is restored. Stabilizing and reducing the mass of the radionuclides and toxic metals contained in such wastes would facilitate their disposal.

The ability of microorganisms which are ubiquitous throughout nature to bring about transformation of organic and inorganic compounds in radioactive wastes has been recognized[1]. Unlike organic contaminants, metals cannot be destroyed, but must be either removed or converted to a stable form. Radionuclides and toxic metals in wastes may be present initially in soluble form or, after disposal, may be converted to a soluble form by chemical or microbiological processes. The key microbiological reactions include (i) oxidation/reduction; (ii) change in pH and Eh which affects the valence state and solubility of the metal; (iii) production of sequestering agents; and (iv) bioaccumulation. All of these processes can mobilize or stabilize metals in the environment. The types of radionuclides and toxic metals commonly found in wastes include uranium, thorium, plutonium, cadmium, cesium, chromium, cobalt, copper, lead, manganese, nickel, strontium, and zinc. They are present in various forms such as elemental, oxide, ionic, inorganic complex, and organic complex. Biotransformation of the various forms of the metal depends upon the type of microbes present, the

Chemical Pretreatment of Nuclear Waste for Disposal, Edited by
W.W. Schulz and E.P. Horwitz, Plenum Press, New York, 1995

presence of suitable electron donors and acceptors, and environmental factors such as pH, temperature, and moisture.

Basic studies at Brookhaven National Laboratory (BNL) dealing with the mechanisms of microbiological transformations of radionuclides and toxic metals have resulted in the development of two novel processes for treating radioactive wastes. One process uses anaerobic bacteria to stabilize the radionuclides and toxic metals in the waste with a concurrent reduction in volume due to the dissolution and removal of nontoxic elements in the waste. In the second process, the toxic metals are removed from the waste by citric acid extraction and the metals and radionuclides in the extract are recovered by biodegradation followed by photodegradation.

EXPERIMENTAL

Waste Sample

A sludge sample was collected from the West End Treatment Facility, at the U.S. Department of Energy, Oak Ridge Y-12 Plant, Oak Ridge, TN. The sludge was generated from a uranium process waste stream after biodenitrification of nitric acid uranium waste. Several million gallons of the sludge is in storage awaiting disposal. Table 1 gives the chemical characteristics of the sludge sample. It contains a small amount of organic carbon and nitrogen, but is high in sulfate concentration and ash content. Sulfate in the sludge resulted from adding sulfuric acid and ferric sulfate during waste treatment. The sludge contains varying levels of major elements, such as aluminum, calcium, iron, magnesium, potassium, and sodium; and toxic metals of arsenic, cadmium, chromium, cobalt, copper, lead, manganese, mercury, nickel, selenium, uranium, and zinc. The concentrations of calcium, nickel, and uranium in the sludge are high.

The mineralogical association of cadmium, chromium, copper, manganese, nickel, lead, uranium, and zinc in the sludge was determined by a selective extraction procedure[2,3]. It involved the determination of water soluble, exchangeable, carbonate, iron-manganese oxide, organic, inert, and residual fractions. For example, uranium was associated with the exchangeable (400 mg), carbonate (1600 mg), iron oxide (216 mg), organic (516 mg), and inert (80 mg/g dry wt) fractions (Figure 1). Nearly half the total uranium in the sludge was associated with the carbonate fraction.

Stabilization and Volume Reduction by Anaerobic Bacterial Treatment

The sludge was treated with anaerobic bacteria to release a large fraction of the waste solids into solution. The radionuclides and toxic metals were converted to a more concentrated and stable form. At the same time, the volume and mass were reduced. In this process, the unique metabolic capabilities of dual-action anaerobic bacteria were exploited to solubilize and precipitate radionuclides[3,4]. The non-hazardous materials in the solid phase such as calcium, potassium, sodium, magnesium, and iron were solubilized and easily removed from the waste, thus reducing its volume. The re-mobilized radionuclides and toxic metals were stabilized by precipitation and redistributed with stable mineralogical fractions of the waste (Figure 2).

Five grams of sludge in 160 ml acid washed sterile serum bottles with and without nutrients were incubated anaerobically in the presence of N_2 atmosphere. One hundred milliliters of deionized water or deionized water containing glucose and NH_4Cl were pre-reduced by boiling for 15 minutes while purging with N_2 (99.99% purity), and then added to the sludge sample; the serum bottles were sealed with butyl

Table 1. Characterization of Oak Ridge site sludge sample.

Constituents	Concentration
Physical, %	
moisture	56.7 ± 0.4*
ash	65.3 ± 0.00
Chemical, % dry weight	
carbon	1.35 ± 0.00
nitrogen	0.12 ± 0.00
sulfate sulfur	2.57 ± 0.05
Major metals, % dry weight	
aluminum	5.21 ± 0.02
calcium	24.1 ± 1.2
iron	0.50 ± 0.01
magnesium	1.30 ± 0.09
potassium	0.05 ± 0.00
sodium	1.87 ± 0.04
Toxic metals, μg/g dry weight	
arsenic	1.3 ± 0.0
cadmium	93.5 ± 0.3
chromium	396 ± 9
cobalt	38.7 ± 10.5
copper	371 ± 1
lead	267 ± 18
manganese	244 ± 4
mercury	10.6 ± 1.1
nickel	1260 ± 5
selenium	<1.0
uranium	2700 ± 200
zinc	1000 ± 30

* ± 1 Standard error of the mean.

rubber stoppers. The samples were inoculated with the _Clostridium_ sp. ATCC #55102. The treatments consisted of (i) sludge sample plus deionized water (unamended); (ii) sludge sample plus deionized water containing 0.5% glucose and 0.015% NH_4Cl (amended); and (iii) control (autoclaved) sludge sample with deionized water. All samples were incubated at 24°C, in triplicate, except the control samples, which were incubated in duplicate. Periodically, unamended, amended, and the control samples were analyzed for the production of total gas, CO_2, H_2, and CH_4. An aliquot of the sample was removed and filtered through a 0.22 μm Millex filter. A portion of the filtered aliquot was analyzed for anions by ion chromatography using a conductivity detector; for organic acids by HPLC, using a UV detector at 210 nm; and for alcohols and glucose by HPLC, using a refractive index detector. Another portion of the aliquot was acidified with HNO_3, and analyzed for uranium and metals.

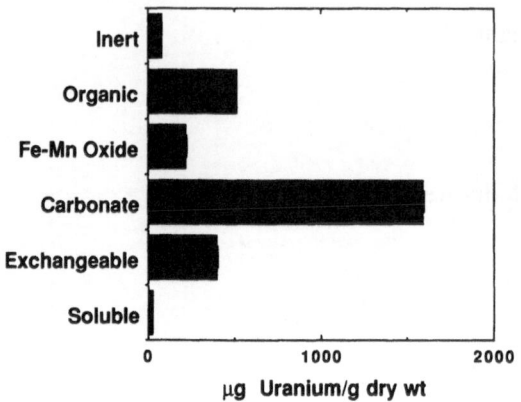

Figure 1. Mineralogical association of uranium in sludge. Uranium was predominantly associated with the carbonate fraction.

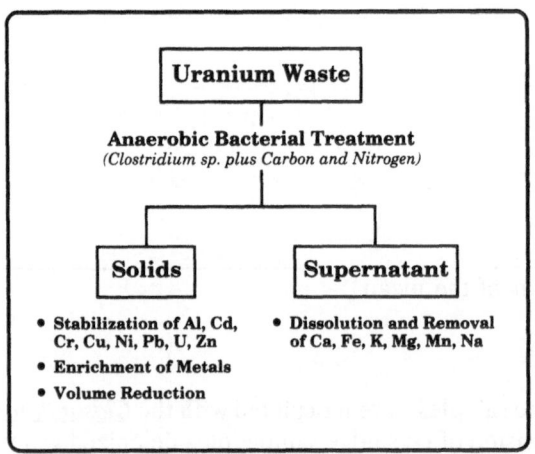

Figure 2. Stabilization and volume reduction of uranium waste. Sludge treated with _Clostridium_ sp. showed stabilization and enrichment of uranium and toxic metals.

Table 2 shows the results of anaerobic microbial activity in the sludge. The unamended samples showed neither significant microbial activity nor production of organic acid metabolic products. However, the amended samples showed an increase in total gas, CO_2, H_2, CH_4, and organic acids; its pH was lowered by about 2.5 units. This change was due to the production of organic acid metabolites from glucose fermentation. The organic acids consisted of acetic, butyric, propionic, formic, pyruvic, lactic, isobutyric, valeric, and isocaproic acids. A significant amount of gas was produced due to glucose fermentation by anaerobic bacteria, as well as the dissolution of $CaCO_3$ in the sludge by the organic acids. A decrease in sulfate concentration was observed only in amended samples, but the formation of sulfides (blackening of waste) was not evident.

Dissolution and Precipitation of Metals. The dissolution of metals due to anaerobic bacterial activity are shown in Table 3. The unamended sludge showed little dissolution of calcium (0.07%), magnesium (2%), nickel (1%), and zinc (0.1%). In the amended samples, increased concentrations of calcium (20%), iron (19%) magnesium (42%), manganese (8%), and lesser amounts of nickel (5%) and zinc (1%) were found in solution, but aluminum, cadmium, chromium, copper, lead, and uranium were not detected in solution. Analysis of the remaining solids showed enrichment of aluminum, cadmium, chromium, copper, nickel, lead, uranium, and zinc (Figure 3). These metals were concentrated in the solids after anaerobic microbial activity.

Mineralogical Association of Metals Before and After Microbial Action. The mineralogical associations of the toxic metals in sludge showed changes after microbial action (Figure 4). Cadmium associated with the carbonate fraction decreased, while the oxide, organic, and inert fractions showed an increase after microbial action. Chromium was present in the carbonate, oxide, organic, and inert fractions and all the fractions showed an increase in concentration of chromium after treatment with anaerobic bacteria. Copper was associated with the organic fraction and a small amount with the oxide and inert fractions. After microbial activity, the concentrations of copper in the organic and inert fractions increased. Most of the manganese was associated with the carbonate fraction. Enrichment of Mn in the carbonate and oxide fractions were observed after microbial activity. The bulk of the nickel was associated with the carbonate and oxide fractions and a small fraction with the organic and inert phase. An increase in concentration of nickel in the oxide, organic, and inert fractions, and a decrease in concentration in the soluble and the carbonate fractions was observed after microbial activity. Lead was predominantly associated with the carbonate and the oxide fractions, with only a small amount present in the organic fraction. The concentration of lead in all three fractions increased as a result of microbial activity. Zinc was present in association with the carbonate, oxide, organic, and inert fractions. After microbial activity the concentration of zinc increased in the oxide, organic, and inert fractions and decreased in the carbonate fraction. Uranium was predominantly associated with the carbonate fraction and to a lesser extent with the oxide, organic, and inert fractions. The concentration of uranium associated with the carbonate, oxide, and inert fractions increased after microbial activity. Figures 5 and 6 show the theoretical vs. observed uranium in solution on the basis of dissolution of calcium and the association of uranium with the carbonate fraction. However, uranium was not detected in solution.

Weight Loss. Substantial amounts of calcium, iron, potassium, magnesium, manganese, and sodium were solubilized from the waste, resulting in a reduction in mass; the net reduction due to the anaerobic bacteria in these batch studies was

Table 2. Anaerobic microbial activity in sludge.

Treatment	pH	Gas Produced (ml)	CO_2	H_2 mmoles	CH_4
Control (Unamended)	8.89 ± 0.03	0.74 ± 0.15	0.965 ± 0.090	0.002 ± 0.000	0.001 ± 0.000
Treated (Amended)	6.40 ± 0.0	52.4 ± 4.6	5.74 ± 0.75	0.771 ± 0.16	0.006 ± 0.000

Notes: Samples were analyzed after 35 days incubation.
Error bar represent ± 1SEM.
Organic by-products detected: ethanol, formic, acetic, propionic, butyric, isobutyric, valeric, isocaproic, lactic, and pyruvic acids.

Table 3. Dissolution of metals from sludge by anaerobic microbial activity.

Treatment	Al	Ca	Cd	Co	Cr	Cu	Fe	K	Mg	Mn	Na	Ni	Pb	U	Zn
								μ gdw^{-1}							
Control (Unamended)	<25	180±19 (0.07)	<1	ND	<2	<1	<5	345±66	240±2 (2)	<1	5240±0	17.6±0.6 (1)	<2	<5	1.24±0.24 (0.12)
Treated (Umended)	<25	49000±5300 (20)	<1	2.53±0.47	<2	<1	966±8 (19)	419±23 (8)	5400±60 (42)	19.9±0.7 (8)	5540±2	69.9±1.4 (5)	<2	<5	10.8±3 (1)

Notes: ( ) = % of total metal.
< values reported for each element are below the detection limit; values reported in parenthesis are % of total metal.
Error bars represent 1 ± SEM.

~ 15%. An additional reduction in waste volume can be achieved by optimizing the process using a continuous culture treatment system to solubilize and remove the bulk of non-toxic waste components, particularly calcium, potassium, iron, magnesium, and manganese.

This biotreatment process can be applied to wastes generated from defense, energy, and industrial wastes containing radioactive elements and toxic metals. The microbial treatment of such mixed wastes chemically converts the radionuclides and metals to a more stable form. Reducing the mass of the wastes means that more material can be stored or disposed of, can be handled easier, and can be transported or chemically treated more easily. Changing the radionuclides and toxic metals to a more stable form allows the material to be, chemically processed more readily or disposed of in shallow or deep geological formation where the material is less susceptible to chemical change or transport. Microbial treatment results in a waste that is less susceptible to environmental factors and can be disposed of more economically.

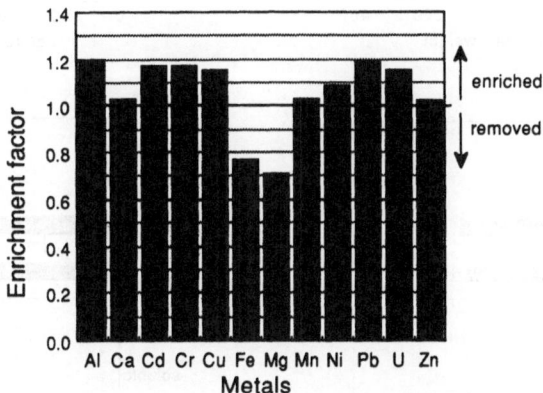

Figure 3. Enrichment of metals in sludge after anaerobic microbial treatment. [Enrichment factor is the mean concentration in the sludge divided by that after microbial activity. A ratio of 1 indicates no action a ratio > 1 shows enrichment, and < 1 shows removal from the sludge.]

Removal and Recovery of Radionuclides and Toxic Metals

An alternative treatment method for radionuclide or toxic metal-contaminated materials, soils, sediments, and wastes involves extracting the metals and radionuclides with citric acid as citrate complexes[5,6]. The extract then is treated with <u>Pseudomonas fluorescens</u> ATCC No. 55241 to recover the toxic metals. The supernatant containing

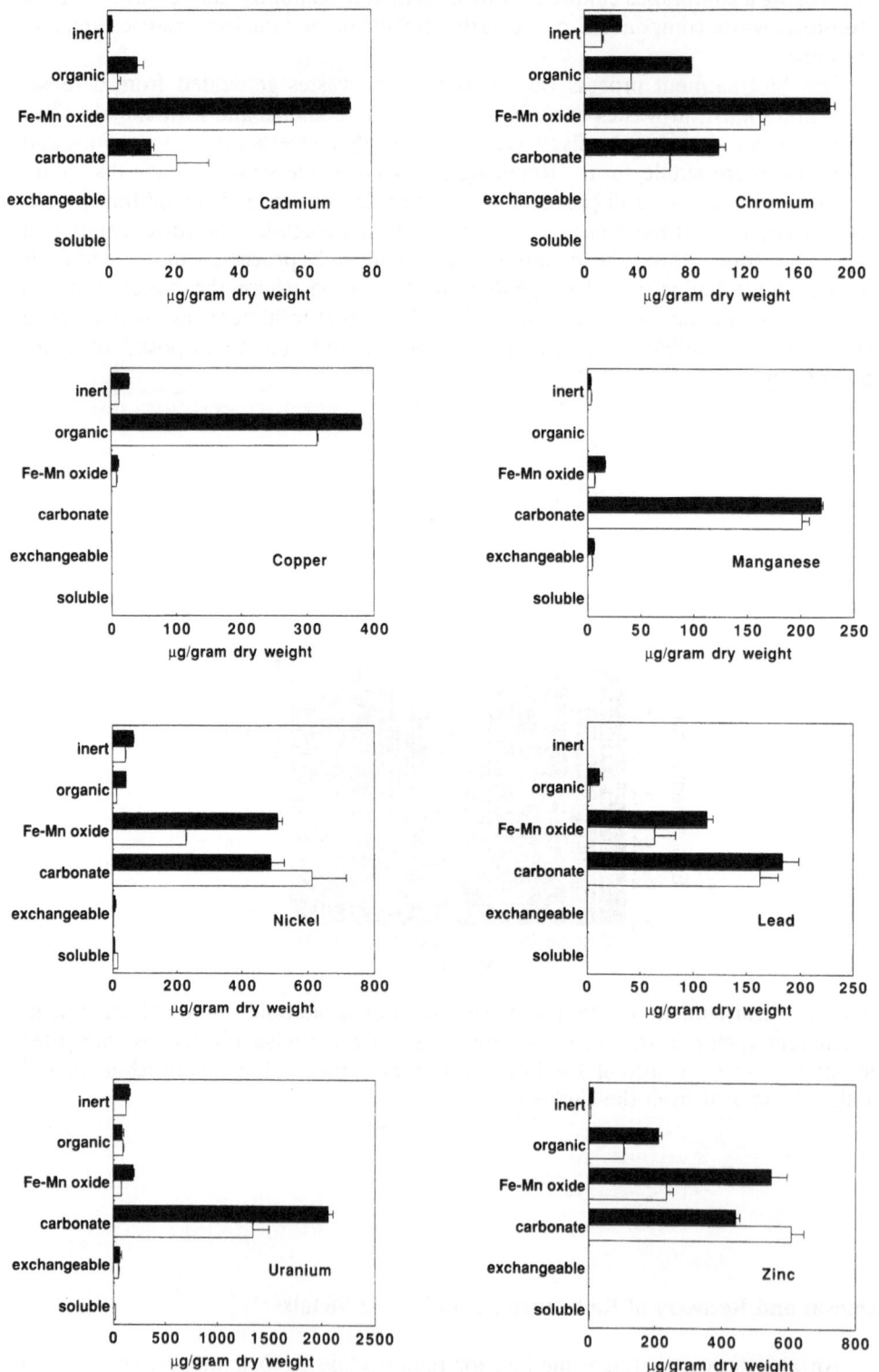

Figure 4. Mineralogical association of uranium and toxic metals in sludge before (□) and after (■) microbiological activity. Error bars represent 1 ± SEM.

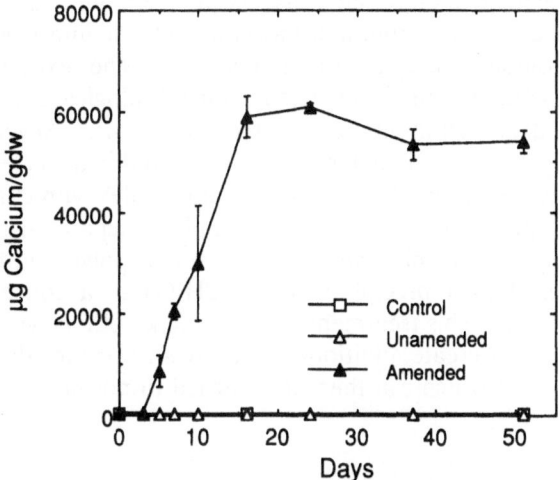

Figure 5. Dissolution of calcium from amended sludge. A substantial amount of calcium was solubilized due to indirect action of the anaerobic bacteria; there was no dissolution in the control and unamended samples.

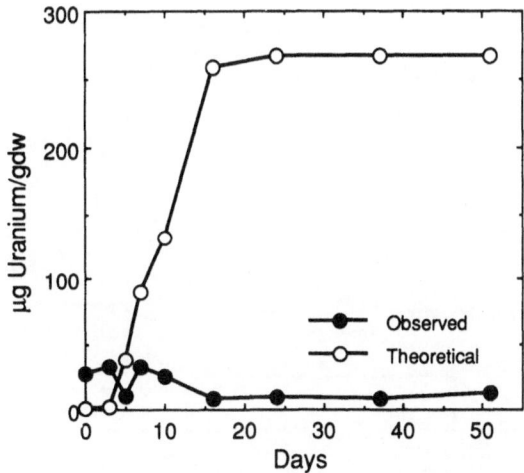

Figure 6. Theoretical and observed concentration of uranium in solution. Uranium associated with the carbonate fraction was released due to dissolution of calcium, and subsequently reduced to a lower oxidation state due to direct enzymatic action of the bacteria.

the uranium-citrate complex is subjected to photolysis to degrade the complex and recover the uranium in a concentrated form (Figure 7).

Citric acid, a naturally occurring organic complexing agent, forms different types of complexes with the transition metals and actinides, and may involve the formation of bidentate, tridentate, binuclear, or polynuclear complex species (Figure 8). Citric acid is used to extract metals such as barium, chromium, nickel, zinc, and radionuclides, such as cobalt, strontium, thorium, and uranium from solid wastes by generating water-soluble, metal-citrate complexes. The extract containing the radionuclide/metal complex then is subjected to microbiological degradation, followed by photochemical degradation under aerobic conditions. Several metal citrate complexes are biodegraded, and the metals are recovered in a concentrated form with the bacterial biomass. Uranium forms a binuclear complex with citric acid and is not biodegraded. The supernatant containing this complex is separated and upon exposure to light, rapidly degrades forming an insoluble, stable polymeric form of uranium. Uranium is recovered as a precipitate ($UO_3 \cdot 2H_2O$) in a concentrated form for recycling or for disposal. This treatment process, unlike others which use alkaline or acidic reagents, does not create additional hazardous wastes for disposal, and causes little damage to the soil which can then be returned to normal use.

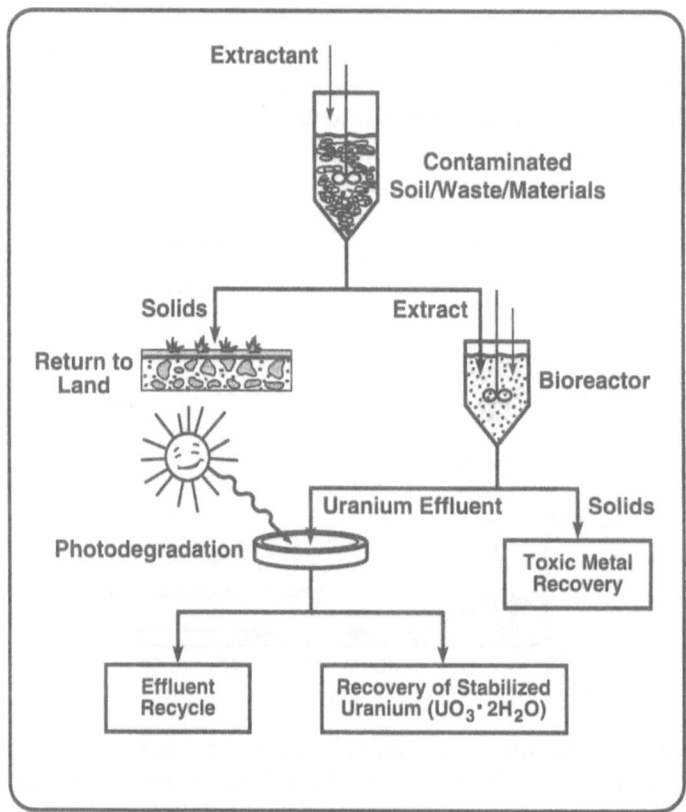

Figure 7. Citric acid treatment process. Flow chart showing the extraction and recovery of uranium and toxic metals from waste by biodegradation/photodegradation processes.

Previous large-scale methods devised to deal with the problems of contaminated materials and soils used corrosive reagents, such as hot sulfuric or hydrochloric acids, and oxidizing agents, such as sodium hypochlorite, to extract the metals. However, these methods also cause irreparable damage to the soil and generate secondary waste streams which create further problems of hazardous waste disposal.

Many chelating agents used in decontamination undergo little degradation by microorganisms[1,7]. Biodegradation of these metal chelates should cause the precipitation of released ions as water-insoluble hydroxides, oxides, or salts, thereby retarding the migration of metals. Recently, we reported that the type of complex formed between the metal and citric acid plays an important role in determining its biodegradability[8,9]. The presence of the free hydroxyl group of citric acid is the key determinant in effecting biodegradation of the metal complex. For example, Ca, Fe(III), and Ni formed mononuclear bidentate complexes and were readily biodegraded, whereas, Cd, Cu, Fe(II), and Pb formed mononuclear tridentate complexes, and U formed a binuclear complex involving the hydroxyl group of the citric acid, and were not biodegraded. The lack of degradation of the latter complexes was not due to their toxicity, but was probably limited by the transport and/or metabolism of the complex by the bacteria[10].

Citric acid has been used to decontaminate components of nuclear reactors, but the methods of metal recovery involve ion-exchange, porous DC electrodes, or burning the organics. Citric acid was used to extract plutonium from contaminated soils to determine plutonium uptake by plants[11]. Several metal citrate complexes are readily biodegraded by microorganisms, with recovery of the metal species with the biomass. Uranyl citrate is recalcitrant to biodegradation but upon exposure to visible light undergoes photochemical degradation resulting in the formation of an insoluble, stable polymeric form of uranium identified as $UO_3 \cdot 2H_2O$[12-14].

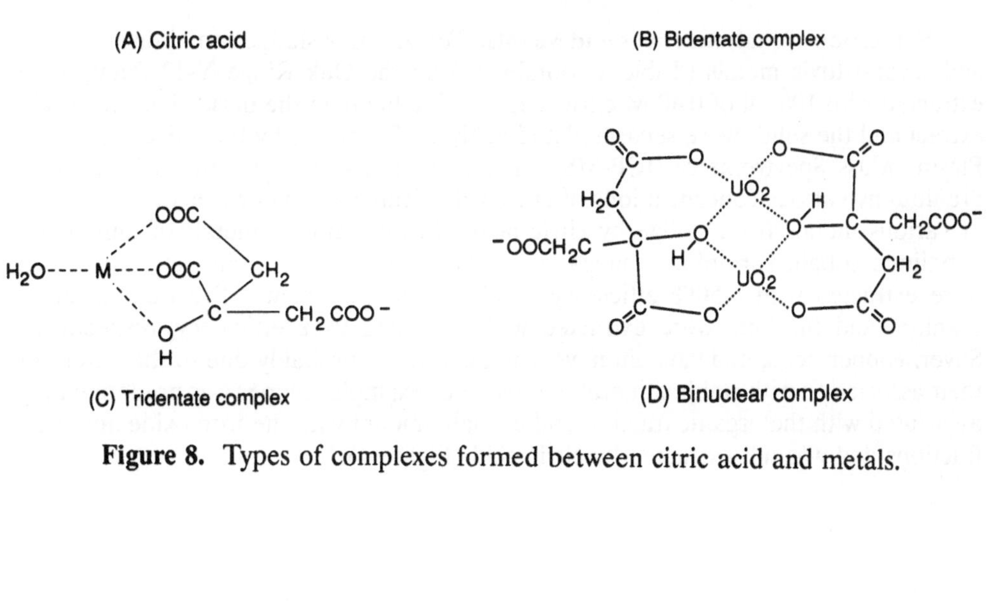

Figure 8. Types of complexes formed between citric acid and metals.

Table 4. Extraction efficiency of metals from sludge by citric acid.

Metal	Total Metals in Sludge (μg.gdw^{-1})		% Metal Extracted
Ag	41	± 30	2.4
Al	30500	± 500	58.7
Au	1800	± 500	<1
Ba	427	± 25	24.4
Be	5.21	± 0.45	60.1
Cd	66	± 6	9.1
Co	0.7	± 0.3	74.8
Cr	342	± 10	74.6
Cu	329	± 18	1.2
Ga	28.8	± 0.6	25.7
Mg	7510	± 100	89.2
Mn	234	± 3	82.9
Ni	1120	± 10	80.0
Pb	224	± 27	<1
Pd	5.51	± 0.70	49.0
Sb	5.67	± 0.05	68.8
Sn	17.6	± 0.4	93.1
Sr	125	± 5	59.2
Th	3.08	± 0.10	94.2
Ti	922	± 95	28.4
U	2410	± 100	86.8
V	121	± 7	4.1
Zn	839	± 7	59.6
Zr	209	± 4	84.2

Extraction of Radionuclides and Metals. Ten grams of sludge containing uranium and several toxic metals (Table 1) obtained from the Oak Ridge Y-12 Plant, were extracted with 100 ml of 0.40 M citric acid, for five hours in the dark. The citric acid extract and the solids were separated and analyzed for metals by Inductively Coupled Plasma-Mass Spectrometry (ICP-MS). Figure 9 shows the extraction efficiency of uranium by various concentrations of citric acid. Table 4 shows extraction efficiency of various metals from sludge by citric acid. In this sample, metals of aluminum, beryllium, cobalt, chromium, manganese, nickel, antimony, tin, zinc, and zirconium were extracted with >50% efficiency by citric acid treatment. The radionuclides, uranium and thorium, were extracted with 87% and 94% efficiency, respectively. Silver, copper, lead, and vanadium were not extracted, probably due to the nature of their association with stable mineral phases[4]. For example, copper was predominantly associated with the organic fraction and a small amount with the iron oxide and inert fractions and was not extracted by citric acid (Figure 4).

Biodegradation of Metal Citrate Extract. Citric-acid sludge extract was amended with nutrients consisting of 0.1% NH_4Cl, K_2HPO_4, and KH_2PO_4. The pH was adjusted to 6.5 with NaOH, and then the extract (100 ml) was inoculated with 4 ml of an 18-hour old culture of <u>Pseudomonas fluorescens</u> ATCC 55241. The samples were incubated for 118 hours on a shaker at 24°C. The bacterial inoculum was grown in medium containing citric acid, 2 g; NH_4Cl, 1 g; KH_2PO_4, 1 g; K_2HPO_4, 1 g; NaCl, 4 g; $MgSO_4$, 0.2 g; distilled water, 1000 ml; and pH 6.5. All samples were perpared under low light to minimize any photochemical reactions. Periodically, 5 ml aliquots were removed, filtered through a 0.22 μm filter, and analyzed for (i) pH; (ii) citric acid biodegradation by HPLC using UV detector; and (iii) soluble uranium. At the end of incubation, the supernatant and the solids consisting of bacterial biomass and the precipitated metals were separated by centrifugation. The dry weight was determined, and the solids were digested in a mixture of hot nitric and perchloric acids. The supernatant liquid and the digested solids were analyzed for uranium and other metals by ICP-MS. The bacteria degraded citric acid at a rate of 0.5-0.7 mM per hour (Figure 10); there was little change in concentration of uranium suggesting that the uranium citrate complex was not biodegraded. Cobalt, nickel, zinc, and zirconium were present along with the biomass, indicating that their citrate complexes were readily biodegraded.

Photodegradation of Uranium Citrate Extract. After adjusting the pH to 3.5 with HCl the supernate from the biodegradation treatment primarily containing uranium citrate complex was exposed to high out put fluorescent growth lights to degrade the complex and recover uranium (Figure 11). Periodically, samples were withdrawn, filtered through a 0.22 μm filter, and analyzed for uranium, citric acid, and photodegradation products. At the end of the experiment (after 157 hours of exposure to light), the solutions were filtered and analyzed for citric acid degradation products and metals. The uranium precipitated out of solution as a polymer soon after it was exposed to light. After 50 hours, ~ 85% of the uranium was removed from solution.

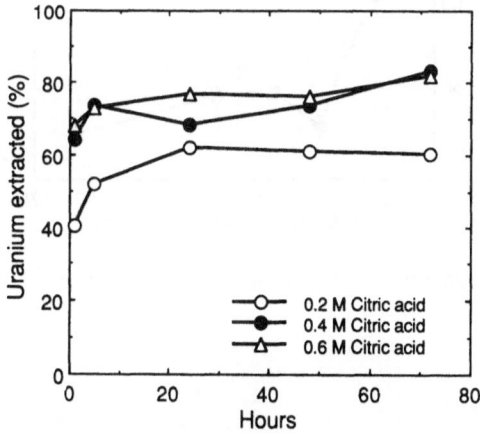

Figure 9. Extraction of uranium from sludge with citric acid.

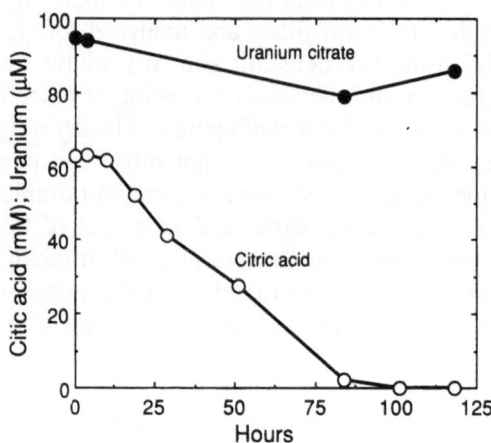

Figure 10. Biodegradation of citric acid sludge extract.

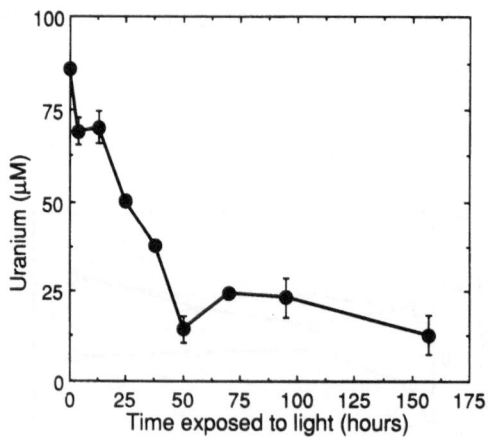

Figure 11. Photodegradation of uranium citrate complex in sludge extract after biodegradation.

Table 5. Effects of biodegradation followed by photodegradation in the treatment of citric acid sludge extract.

Metal	Before Treatment (μM)	% Removal		
		After Bio-Treatment	After Photo-Degradation	Total
Al	7410	92	1	93
Ba	12.0	92	<1	92
Be	3.22	>99	<1	>99
Co	0.87	71	6	77
Cr	60.8	<1	25	25
Ga	0.89	73	16	89
Mn	37.2	98	<1	98
Ni	192	64	1	65
Pd	0.31	64	30	94
Sb	0.36	2	14	16
Sn	1.71	>99	<1	>99
Sr	10.2	98	1	99
Th	0.11	96	3	>99
Ti	80.2	96	<1	96
U	94.8	9	78	87
Zn	86.1	90	5	95
Zr	61.7	97	2	99

The efficiency of removal of various metals from the citric acid sludge extract subjected first to biodegradation, then to photodegradation is shown in Table 5.

Characterization of Uranium Precipitate. The uranium precipitate was identified as ($UO_3 \cdot 2H_2O$) by X-ray Photoelectron Spectroscopy (XPS) and by X-ray Absorption Near-Edge Spectroscopy (XANES) at the National Synchrotron Light Source (NSLS)[14]. The uranium precipitate was quite insoluble at the near neutral pH and soluble in acidic pH (<3.5).

Weight Loss. The solids remaining after extraction with citric acid were washed with deionized water, and dried in an oven overnight at 105°C to determine the weight loss due to citric acid extraction. The extraction of metals from the waste resulted in significant reduction (47%) in weight.

These results show that (i) uranium was extracted from the mixed waste with >85% efficiency using 0.4 M citric acid; (ii) other metals such as chromium, cobalt, manganese, nickel, strontium, thorium, zinc, and zirconium were also extracted from the waste; (iii) the uncomplexed excess citric acid and several metal citrate complexes (Co, Ni, Zn and, Zn) with the exception of binuclear uranium citrate complex, were readily biodegraded by Pseudomonas fluorescens, and the metals were recovered with the bacterial biomass; and (iv) the uranium citrate complex was photodegraded, allowing the uranium to form a polymer which was recovered as a concentrated solid.

SUMMARY

Two microbiological processes are described for treating wastes, soils, and materials containing radionuclides and toxic metals. In one process, anaerobic microbes are used to concentrate, contain, and stabilize the toxic metals and radionuclides in the waste with concurrent reduction in volume. In the second process, the radionuclides and toxic metals are removed from the waste or contaminated material and recovered in a concentrated form for recycling or for disposal.

Stabilization and Volume Reduction by Anaerobic Bacterial Treatment. Stabilization of toxic metals and radionuclides in wastes is accomplished by exploiting the unique metabolic capabilities of the anaerobic bacterium, _Clostridium_ sp. Mixed wastes, sludges, and contaminated soils or sediments are treated with the anaerobic bacteria. In this novel approach, the radionuclides and toxic metals are solubilized by the bacteria directly by enzymatic reductive dissolution or indirectly due to the production of organic acid metabolites. The radionuclides and toxic metals released into solution then are immobilized in the waste matrix by (i) enzymatic/chemical reductive precipitation processes; (ii) redistributed with stable mineral phases in the waste due to readsorption with the reactive surfaces; and (iii) biosorbed by the biomass. The re-mobilized radionuclides and toxic metals are thus stabilized by bacterial action. The iron, calcium, and other non-hazardous materials in the soluble phase are easily removed from waste so reducing its volume. The remaining solids, which contain the hazardous components in a concentrated, stable form, then can be disposed of safely in the subterranean environment. The reduced volume of the waste also results in considerable savings in disposal costs.

Removal and Recovery of Uranium and Toxic Metals from Waste. Removal and recovery of toxic metals and uranium in waste is accomplished aerobically using citric acid, aerobic microorganisms, and photochemical action. Uranium and toxic metals are removed from wastes or contaminated materials or soils by extracting with the complexing agent citric acid. The citric acid extract is subjected to microbiological degradation to recover the toxic metals, followed by photochemical degradation of the uranium citrate complex which is not biodegraded. The toxic metals and uranium are recovered in a concentrated form for recycling or for disposal. This process has significant potential for commercialization because (i) it can be applied to a variety of materials and waste forms; (ii) mixed waste is separated into radioactive and hazardous waste; (iii) uranium is separated from the toxic metals and recovered for recycling or disposal; (iv) it does not generate secondary waste streams; (v) it causes little damage to soil; and (vi) environmentally and economically important metals are removed in a concentrated form. The use of combined chemical, photochemical, and microbiological treatment processes of contaminated materials will be more efficient than present methods and result in considerable savings in clean-up and disposal costs.

ACKNOWLEDGEMENTS

I thank C. J. Dodge, J. B. Gillow, and C. C. Neill for their assistance and contribution to this work. This research was performed under the auspices of the Environmental Sciences Division's Subsurface Science Program, Office of Health and Environmental Research, Office of Energy Research, U. S. Department of Energy, under Contract No. DE-AC02-76CH00016.

REFERENCES

1. Francis, A. J. "Microbial transformation of toxic metals and radionuclides in mixed wastes." *Experientia* (1990) 46, 840-851.

2. Tessier, A., Campbell, P. G. C., and Bisson, M. "Sequential extraction procedure for speciation of particulate trace elements." *Anal. Chem.* (1979) 51, 844-851.

3. Francis, A. J., Dodge, C. J., Gillow, J. B., and Cline, J. E. "Microbial transformation of uranium in wastes." *Radiochimica Acta.* (1991) 52-53, 311-316.

4. Francis, A. J. and Dodge, C. J., and Gillow, J. B. "Microbial stabilization and mass reduction of wastes containing radionuclides and toxic metals." U.S. Patent No. 5,047,152 (1991).

5. Francis, A. J., Dodge, C. J. "Reclamation with recovery of radionuclides and toxic metals from contaminated materials, soils, and wastes." *Technology 2002, NASA Conference Publication 3189*, (1992) Vol. 1, 109-117.

6. Francis, A. J. and Dodge, C. J. "Waste site reclamation with recovery of radionuclides and metals." U.S. Patent No. 5,292,456 (1994).

7. Francis, A. J. "Microbial transformation of low-level radioactive wastes in subsoils" in *Soil Reclamation Processes: Microbiological Analyses and Applications*, R. L. Tate and D. Klein, Editors, Marcel Dekker: New York, (1985) pp. 279-331.

8. Francis, A. J., Dodge, C. J., and Gillow, J. B. "Biodegradation of metal citrate complexes and the implications for toxic metal mobility." *Nature* (1992) 356, 140-142.

9. Francis, A. J. and Dodge, C. J. "Influence of complex structure on the biodegradation of iron-citrate complexes." *Appl. Environ. Microbiol.* (1993) 59, 109-113.

10. Joshi-Topé, G. A. and Francis, A. J. Mechanisms of degradation of metal citrate complexes by Pseudomonas fluorescens. (submitted).

11. Nishita, H., Havg, R. M., and Rutherford, T. "Effect of inorganic and organic compounds on the extractability of ^{239}Pu from an artificially contaminated soil." *J. Environ. Qual.* (1977) 6, 451-455.

12. Adams, A. and Smith T. D. "The formation and photochemical oxidation of uranium(IV) citrate complexes." *J. Chem. Soc.* (1960) 4, 4846-4850.

13. Ohyoshi, A. and Ueno, K. "Studies on actinide elements - (IV): Photochemical reduction of uranyl ion in citric acid solution." *J. Nucl. Chem. Inorg.* (1974) 36, 379-384.

14. Dodge, C. J. and Francis, A. J. Photodegradation of uranium citrate complex with uranium recovery. *Environ. Sci. Technol.* (1994) 28, 1300-1306.

TREATMENT OF HIGH-LEVEL WASTES FROM THE IFR FUEL CYCLE

T. R. Johnson, M. A. Lewis, A. E. Newman, and J. J. Laidler

Chemical Technology Division
Argonne National Laboratory
9700 South Cass Avenue
Argonne, IL 60439

INTRODUCTION

The Integral Fast Reactor (IFR)[1] is an advanced power reactor concept sponsored under the U.S. Department of Energy's reactor development program.[2] It promises to have important advantages over present reactors in its capability to conserve resources and consume the transuranic elements, thus excluding these from its wastes. It can also be used to consume the actinides in other nuclear wastes, such as spent light-water reactor fuels, and the transuranic (TRU) elements from nuclear weapons. The IFR features on-site reprocessing using pyrochemical methods to recover actinides from the U-Pu-Zr core and U-Zr blanket fuels, separate fission products, and produce suitable high-level waste forms. More than 99% of the TRU elements are recovered and separated from fission products by an electrorefining process at about 500°C.[3] Treatment of electrorefiner wastes to recover residual actinides results in the recovery, recycle, and burnup of more than 99.9% of the TRU elements in discharged fuel. The electrorefiner (Figure 1) is a steel vessel that holds a pool of liquid cadmium (which serves as an anode in some operations) and a liquid LiCl-KCl electrolyte. Spent fuel elements are chopped, placed in steel baskets, and immersed in the electrolyte, where the fuel is anodically dissolved from the cladding. A major fraction of the uranium from the fuels is electrochemically deposited on steel cathode rods, and the rest of the uranium and the TRU elements are deposited in liquid cadmium cathodes. A significant fraction of the zirconium is also deposited with uranium on the solid cathodes. These cathodes are retorted to recover salt and cadmium, and melted to produce metal ingots for use in producing new core and blanket fuels.

The fission products, with the exception of tritium, krypton, and xenon accumulate in the electrorefiner during processing. Some of the noble metal fission products remain in the anode baskets with the cladding hulls, but a significant fraction fall into the cadmium pool. The alkali metal, alkaline earth, rare earth, and halide fission products collect in the salt. After many batches of core and blanket fuels are processed, the salt and cadmium in the electrorefiner are treated to recover and recycle residual TRU elements and to concentrate the fission products for conversion to high-level waste forms, which will be placed in a geologic repository.

TREATMENT OF SPENT SALT AND CADMIUM FROM ELECTROREFINER

After an electrorefining campaign during which many batches of core and blanket fuels are processed, operations are performed to prepare for the removal of salt and cadmium containing the accumulated fission products. The first operation entails

Chemical Pretreatment of Nuclear Waste for Disposal, Edited by
W.W. Schulz and E.P. Horwitz, Plenum Press, New York, 1995

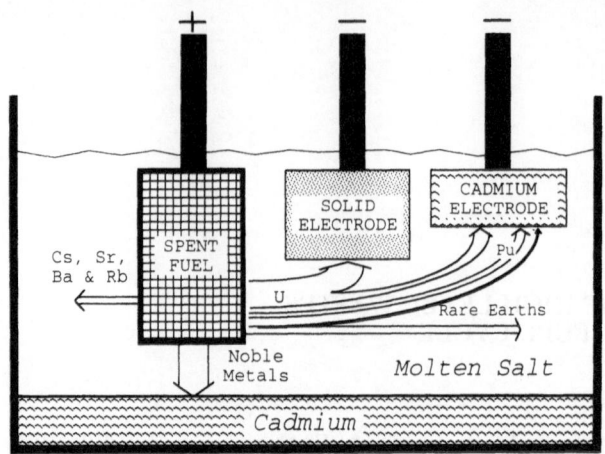

Fig. 1. Separation of Actinides and Fission Products in IFR Electrorefining Process

removing all of the actinides from the cadmium pool and most of the actinides from the salt. These are added to the electrorefiner in the next processing cycle.) Then the salt and cadmium are pumped through a steel filter to remove any insoluble impurities (e.g., oxides, nitrides, and carbides) and undissolved noble metals (Zr, Mo, Tc, Ru, and Rh). The filter element retains some cadmium so that a fraction of the soluble noble metals (Pd, Ag, Sn, and Sb) can be removed from the pool. The filter and then most of the salt are removed from the electrorefiner. The estimated amounts of spent salt and metal, based on processing 500 kg of actinides from core and blanket fuels with an overall average burnup of 60,000 MWd/T, are shown in Table 1.

The sodium in the spent LiCl-KCl salt (see Table 1) arises when the bond sodium in the chopped fuel elements, fed to the electrorefiner, is converted to chloride by the addition of $CdCl_2$. Yttrium, samarium, and europium are included with the alkaline earths, because these rare earths behave like alkaline earths and remain in the salt during the salt treatment steps.

The amounts of alkali metal, alkaline earth, and rare earth fission products in the salt exceed those fed with the 500 kg of actinides, because salt treatment removes only a fraction of these elements and returns the rest to the electrorefiner with recycled salt. The amounts of zirconium and noble metals in the discharged cadmium are the estimated amounts that are not recovered with uranium cathodes or that are not retained in the anode baskets.

Table 1. Spent Salt and Cadmium Discharged from Electrorefiner

Salt	Weight of Chloride (kg)	Metal	Weight of Metal (kg)
LiCl-KCl-NaCl	445	Cd	100
RbCl-CsCl	20	Zr	15
$SrCl_2$-$BaCl_2$[a]	16	Noble Metals	5
(Rare Earth)Cl_3	15		
UCl_3	1	U	<0.1
(TRU)Cl_3	3	TRU	<0.01
Totals	500		120

[a]Includes YCl_3, $SmCl_2$, and $EuCl_2$.

Figure 2 illustrates the process steps used to recover TRU elements, salt, and cadmium for return to the electrorefiner, and to prepare high-level waste forms for the remaining salt and metal.[4] Spent salt from the electrorefiner flows through an extraction train, which consists of several countercurrent stages, where it is contacted with a Cd-2 wt % U solution. This step extracts more than 99% of the TRU elements and leaves more than 70% of the rare earths in the salt. Actinides and rare earths in the product cadmium solution are separated from the cadmium by retorting and returned to the electrorefiner; the cadmium is recycled to the extractor after the addition of uranium.

The product salt from the extractor, which contains fission products, uranium, and very small amounts of TRU elements, is sent to the salt stripper, where it is contacted with a liquid Cd-Li alloy to reduce essentially all of the actinides. With the exception of yttrium, samarium, and europium more than 99% of the rare earth elements are also transferred to the cadmium solution. In addition to those three rare earths, the stripped salt contains nearly all of the alkali metal, alkaline earth, and halide fission products, but only trace amounts of actinides.

The stripped salt passes through a zeolite bed that removes cesium, strontium, and barium by ion exchange. Excess salt is removed from the bed by a final purge with argon gas, but the bed retains some salt occluded in the zeolite cavities and adhering to the zeolite surface. Most (≈90%) of the salt, with its fission product content reduced by 10 to 30%, passes through the zeolite bed and is returned to the electrorefiner. The salt-laden zeolite is then mixed with an aluminosilicate material and hot pressed to form a high-level waste form consisting of salt contained in a dense ceramic matrix, such as sodalite.

The combined metal wastes (including the filter element from treatment of the electrorefiner contents) are retorted to recover cadmium, which is returned to the process. A matrix metal, e.g., a copper alloy, is added during the retorting process to disperse and immobilize the residue, which consists of the noble metal and rare earth fission products, small amounts of uranium, trace amounts of TRU elements, and cladding hulls.

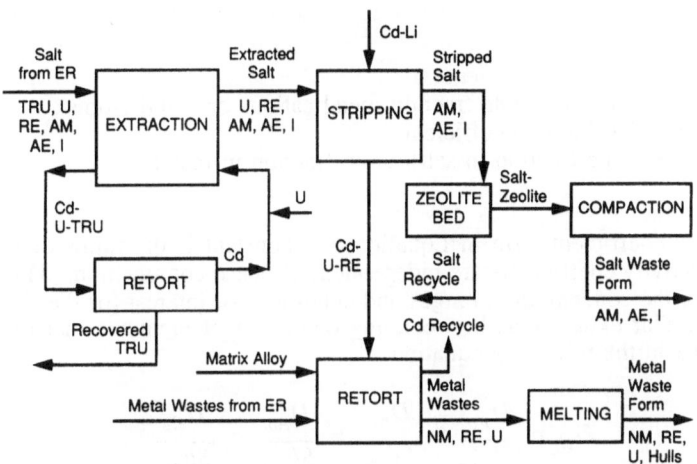

Fig. 2. Reference Process for Treating and Containing and Salt and Metal Waste
(RE = rare earth, AM = alkali metal, AE = alkaline earth, NM = noble metal fission products)

CHEMICAL BASIS OF SALT EXTRACTION AND STRIPPING

Salt extraction and stripping, as well as the electrorefining process itself, are based on the distribution of metallic elements dissolved in cadmium and their chlorides dissolved in a mixture of stable alkali metal and alkaline earth chlorides.[3] Equilibrium among the elements that distribute between the phases is achieved by exchange reactions between cations in the salt and metal atoms in the cadmium. Equilibrium among elements that form trivalent chlorides can be represented by exchange reactions between any pair of elements, for example,

$$UCl_3 + Pu = U + PuCl_3, \qquad (1)$$

$$UCl_3 + Ce = U + CeCl_3. \qquad (2)$$

Uranium, TRU elements, and most rare earth fission products form only trivalent chlorides under IFR process conditions. Samarium and europium form divalent chlorides and tend to behave like alkaline earths in this process. Exchange reactions involving cadmium metal and chlorides of the alkali metals, alkaline earths, and the rare earths yttrium, samarium, and europium can be ignored under IFR process conditions. The equilibrium constant, K, for Reaction 1 is

$$K = \frac{[U\ ]\,[PuCl_3]}{[UCl_3]\,[Pu]} = \exp(-\Delta G^{\circ}/RT), \qquad (3)$$

where [U]   =   activity of uranium in cadmium,
      [Ucl$_3$]   =   activity of Ucl$_3$ in salt,
      ΔG° =   standard free energy change of the reaction,
      R   =   gas constant, and
      T   =   absolute temperature.

This leads to an expression for the plutonium-uranium separation factor,

$$SF_{Pu} = \frac{D_{Pu}}{D_U} = K \left\{ \frac{(\gamma Pu)\,(\gamma UCl_3)}{(\gamma PuCl_3)\,(\gamma U)} \right\}, \qquad (4)$$

where γ = activity coefficient
             = (activity)/(mole fraction metal cations or metal atoms),
and D   = distribution coefficient
             = (mole fraction in salt)/(atom fraction in metal).

The activity coefficient term in Equation 4 is constant if the ratios of the salt- and metal-phase activity coefficients are independent of phase composition. This appears to be the case for the concentration ranges and conditions of interest for the IFR fuel cycle processes. Similar expressions between any other pair of elements that form trivalent chlorides result in the following equation:

$$D_U = \frac{D_{Pu}}{SF_{Pu}} = \frac{D_{Np}}{SF_{Np}} = \frac{D_{Am}}{SF_{Am}} = \frac{D_{Ce}}{SF_{Ce}} \,... \qquad (5)$$

where the separation factors are relative to uranium. (The choice of uranium as the reference element is arbitrary; any trivalent element could be chosen.) This relationship can be extended to an element (A) that forms a monovalent or divalent chloride if the separation factor is defined as:

$$SF_A = \left[(D_A)^{3/n} \right] / D_U, \qquad (6)$$

where n = valence of the chloride formed by element A. Process calculations are greatly simplified through the use of these separation factors because they have been found to be constant over wide ranges of redox conditions in the IFR salt/cadmium systems.

Although separation factors can, in principle, be calculated from thermodynamic properties, it has been necessary to measure the actinide and rare earth separation factors. Distribution data for uranium, TRU elements, and some rare earths, which were collected by several researchers during the development of the electrorefining and waste treatment processes, were analyzed by Koyama et al.[5] Results for the LiCl-Kcl eutectic at 500°C are shown in Figure 3. The distribution coefficients in this figure are the ratios of weight fractions rather than mole fractions.

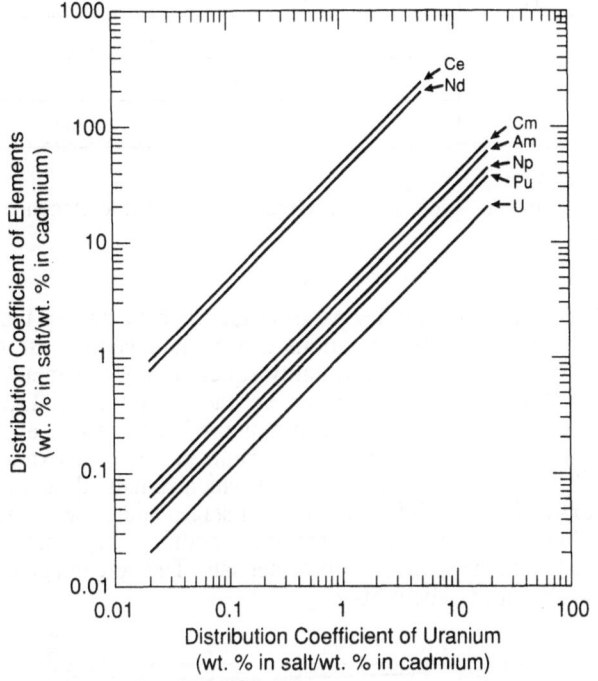

Fig. 3. Distribution Coefficients of TRU Elements and Some Rare Earths
in Alkali-Metal Rich Salt/Cadmium Systems at 500°C

Ackerman and Settle measured distribution coefficients for several rare earths and related these data to the actinide coefficients.[6,7] To overcome the experimental difficulties of determining separation factors larger than about 50, they made measurements with several pairs of elements that progressively favored the salt phase more strongly. Separation factors, as defined by Equation 6, that were derived from these data are listed in Table 2. Samarium and europium form divalent chlorides in this salt/cadmium system. Measurements by Ackerman and Settle[7] of the distribution of several alkali metals and alkaline earths have confirmed that sodium, potassium, cesium, strontium, and barium will remain almost completely in the salt phase during the IFR fuel recovery process.

The large separation factors between rare earths and actinides show that these element groups can be effectively separated by liquid salt-metal extractions. The small separation factors among the actinides are a distinct advantage for the IFR concept, because it is difficult to separate uranium and the minor actinides from plutonium to produce high purity plutonium in an IFR facility.

The basis of the extraction process can be illustrated by considering the following. When a salt containing uranium, TRU elements, and rare earths is contacted with cadmium containing only uranium, the elements will redistribute by means of exchange reactions to satisfy the equilibrium relationship of Equation 5. The net result is that TRU elements and rare earths in the salt exchange with uranium in the metal, but a larger fraction of the TRU elements than rare earths transfers. Uranium replaces TRU elements in the salt and eventually becomes part of the high-level metal waste. This uranium loss, which is less than 1% of the actinides fed to the electrorefiner, is not a serious penalty because it is depleted uranium.

Table 2. Separation Factors of Some Actinides and Rare Earths (LiCl-KCl eutectic/Cd at 500°C)

Element	Separation Factor[a]	Ref.	Element	Separation Factor[a]	Ref.
Np	2.12	5	Ce	4.9E+1	7
Pu	1.88	5	La	1.3E+2	7
Am	3.08	5	Gd	1.5E+2	7
Cm	3.52	5	Y	6.0E+3	7
Pr	43.1	7	Sm	$<10^{10}$	(b)
Nd	44.0	6	Eu	$<10^{10}$	(b)

[a]Defined by Equation 6.
[b]Estimate. Personal communication. J. P. Ackerman and J. L. Settle, Argonne National Laboratory, 1992.

Calculations[8] using the measured separation factors have shown that adequate TRU element recoveries and rare earth separations can be achieved by use of four to seven countercurrent, equilibrium stages with roughly equal volumes of metal and salt phases. Typical results are shown in Figure 4, where the fractions of TRU elements and "extractable" rare earths remaining in the product salt are shown as functions of the metal/salt weight ratio and the number of equilibrium stages. The extractable rare earths do not include yttrium, samarium, and europium which remain almost completely in the salt phase during extraction. With four theoretical stages and a metal/salt weight ratio of 4.0, 99.8% of the plutonium and 99.0% of the minor actinides neptunium, americium, and curium are recovered from the spent electrorefiner salt. The percentage of extractable rare earths remaining in the salt is about 80%.

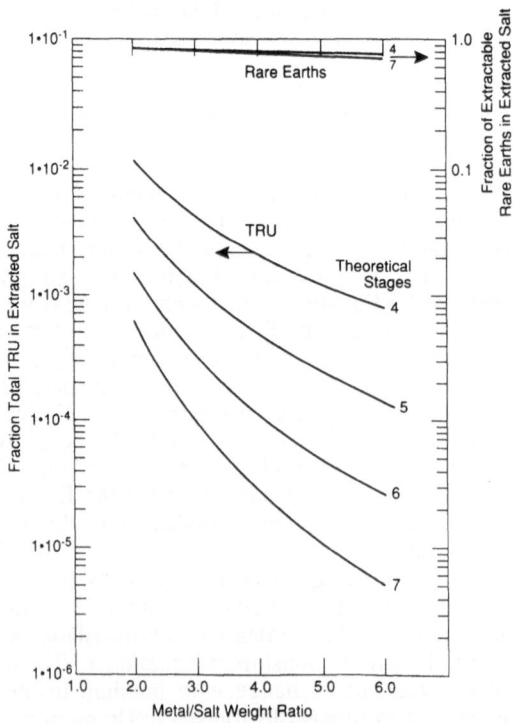

Fig. 4. Calculated Separation of TRU Elements and Rare Earths for
Countercurrent Salt-Metal Extraction
System: LiCl-Kcl/Cd at 500°C

The chemistry of the salt stripper is based on the same distribution behavior of actinides and rare earths between salt and cadmium as the extractor, but the system is made highly reducing by the addition of sufficient lithium metal to reduce essentially all of the actinides and extractable rare earths from the salt and produce a final metal solution with 0.01 to 0.1 wt % Li. The uranium distribution coefficient at this condition is estimated to be less than 1×10^{-6}. As a result, the TRU element distribution coefficients calculated from the measured separation factors are less than 4×10^{-6}, and the coefficients for the extractable rare earths range from roughly 4×10^{-5} for praseodymium and neodymium up to 3×10^{-4} for gadolinium.

The production of low-TRU element salt by this stripping technique has been confirmed in several tests using one-kilogram batches of salt discharged directly from a laboratory-scale electrorefiner.[9] The salt contained 0.2 wt % U and 1 wt % total rare earths yttrium, cerium, neodymium, and samarium which are approximately the expected levels in salt from the IFR extraction step, and 1.2 wt % Pu and 30 ppm Am, which exceed the expected levels. After a single contact with a cadmium solution having a final lithium content of about 0.1 wt %, the uranium, cerium, and neodymium contents of the salts were below detectable limits of 0.01 wt %. The plutonium activity in the salt was less than 1 nCi/g, corresponding to a distribution coefficient of less than 1×10^{-6}, but the americium coefficient was about 1×10^{-3}, not the less than 2×10^{-6} calculated from the plutonium coefficient. There is some experimental evidence for the formation of $AmCl_2$ under very reducing conditions. This could explain the higher-than-expected americium content in the stripped salt. Even with an americium distribution coefficient of 1×10^{-3}, the stripped salt in an IFR process would have a total alpha activity of less than 10 nCi/g.

During these experiments, it was found that direct contact of the salt with a Cd-Li alloy produced intermetallic compounds of cadmium with actinides and rare earths that collected on the crucible walls and did not dissolve readily in the metal phase. Evidently, the chloride reduction rate exceeded the rate at which the metal reduction product could be dissolved by the cadmium. The result was the formation at the salt-metal interface of solid intermetallic compounds that were swept to the sides of the vessel by the agitator. In large-scale tests,[9] inserting inverted cups of Cd-30 wt % Li alloy into the liquid cadmium phase was found to be an effective method for achieving complete dissolution of reduced actinides and rare earths.

ION EXCHANGE WITH ZEOLITES IN MOLTEN SALTS

As shown in Figure 2, the stripped salt passes through a bed of zeolites that preferentially removes a fraction of the strontium, cesium, barium, and rare earths from the salt by cation exchange. Most of the salt passes through the bed and is returned to the electrorefiner, but some is retained in the bed as salt occluded in the molecular cages of the zeolite and as salt adhering to the surfaces of zeolite particles.

Preliminary experiments with various zeolites showed that zeolite A has desirable ion exchange and salt occlusion properties for removing alkali metal and alkaline earth fission products from molten LiCl-KCl solutions.[10] Ion exchange between the molten salt and zeolite A was demonstrated in two batch experiments through changes in the composition of the zeolite before and after equilibration with LiCl-KCl eutectic salt containing $SrCl_2$, CsCl, $BaCl_2$, and NaI. Initially, the zeolite contained 16.2 wt % sodium and no strontium, cesium, barium, or iodine.

After equilibration at 400°C in the first batch experiment, the zeolite was washed with water to remove the surface or non-occluded salt. Table 3 gives the analyzed amounts of strontium, cesium, barium, and iodine in the washed, salt-occluded zeolite, as well as the calculated amount (from a mass balance) in the salt. In the second batch experiment, more than 80% of the molten salt was separated from the zeolite by pressure filtration. Table 3 also gives the amounts of strontium, cesium, barium, and iodine as determined by analysis of the separated salt and material balance calculation for the salt-occluded zeolite.

The data in Table 3 demonstrate that ion exchange occurred between molten salt and zeolite A. The salt-occluded zeolite contained significant amounts of strontium, cesium, and barium while the concentrations of these elements in the salt decreased. More complete analyses of the salt-occluded zeolite showed that between 10 and 12 salt molecules were occluded in the zeolite cavity. It was not possible to distinguish between the occluded cations and the ion exchanged cations. The iodide content of the molten salt

Table 3. Compositions of Selected Elements in Salt and Zeolite from Batch Experiments

Component	Initial Salt (wt %)	Batch #1 (Salt/Zeolite,5.2 g/g)		Batch #2 (Salt/Zeolite,6.1 g/g)	
		Compositions after Equilibration (wt %)			
		Calc. Molten Salt[a]	Exp. Salt-Zeolite[b]	Exp. Molten Salt[c]	Calc. Salt-Zeolite[a]
Sr	0.41	0.021	1.5	0.034	1.7
Cs	4.1	2.5	6.7	2.9	6.2
Ba	1.40	0.24	4.5	0.27	5.1
I	0.40	0.39	0.13	0.36	0.1

[a]Calculated from material balance.
[b]Analyses of salt-occluded zeolite.
[c]Analyses of filtered salt samples.

phase was lowered only slightly by contact with the zeolite, indicating that the salt in the zeolite cavities was enriched in iodide compared to the molten salt. After both experiments, the sodium content of the zeolite had decreased from about 16 wt % to less than 1 wt %, while the sodium content of the salt had increased.

Other data for cesium and strontium adsorption on zeolite A at 450°C are shown in Figure 5. The ordinate is the cation equivalent sorbed per mole of zeolite with a unit cell represented by $Na_{12}[(AlO_2)(SiO_2)]_{12}$. These data were obtained by adding increments of cesium or strontium chloride to a mixture of LiCl-KCl and zeolite particles, stirring for several hours, and extracting filtered samples from the salt solution. The amount of the cation adsorbed by ion exchange with sodium in the zeolite was determined from a mass balance.

The ion exchange step in the process flowsheet has been demonstrated in experiments in which zeolite A (sodium form) was mixed with 30 g of LiCl-KCl solution containing 0.44 wt % Sr, 3.9% Cs, 1.3% Ba, and 0.42% I. With a salt/zeolite weight ratio of 6/1 and a temperature of 400°C, these values were reduced to 0.06 wt % Sr, 3.0% Cs, 0.37% Ba, and 0.39% I. After passing through either a 5- or 20-μm steel filter, the salt

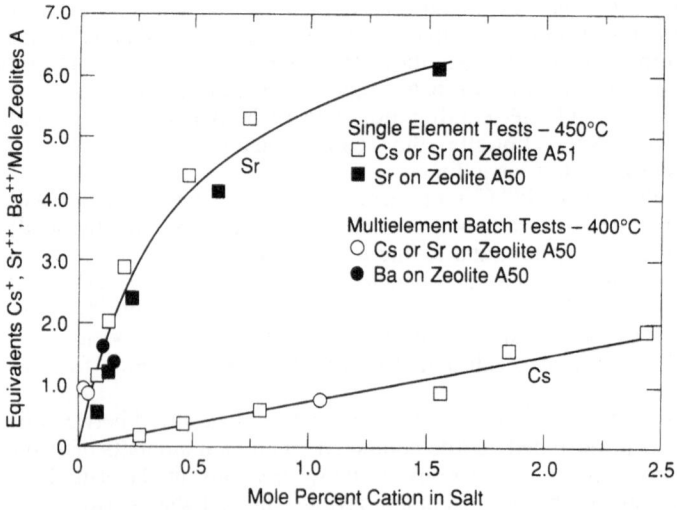

Fig. 5. Adsorption of Cesium, Strontium, and Barium
on Zeolite A from LiCl-KCl Eutectic

contained less than 10 ppm Al and 50 ppm Si. A filtered salt with these low impurity contents could be returned to the electrorefiner. The adsorption data for strontium, cesium, and barium obtained in the filtration tests are plotted in Figure 5 along with the data obtained with only strontium or cesium present. Although the temperatures were different, there is fair agreement between the two sets of data.

Process calculations based on the above results show that the removals of strontium, cesium, barium, and iodine obtained by ion exchange and salt occlusion provide adequate decontamination of the salt so that the effluent salt from a zeolite column can be returned to the electrorefiner. In addition, the fission product concentration by the zeolite is more than sufficient to minimize waste form volume.

SALT WASTE FORM DEVELOPMENT

The salt remaining in the zeolite column is expected to contain negligible amounts of actinides and trivalent rare earths. However, the salt-zeolite mixture will be a high-level waste because it contains large amounts of several relatively short-lived nuclides, namely, Sr-Y-90, Cs-Ba-137, Sm-151, Eu-154, and Eu-155. The heat produced by these radionuclides must be considered in the design of a waste form. A form made by directly compacting the salt-zeolite mixture would have an excessively high fission product heat rate. Diluting the salt-zeolite mixture with an aluminosilicate matrix so that the waste form contains about 10% salt yields a high-level waste form with a heat rate suitable for disposal in a geologic repository. The presence of several long-lived nuclides (I-129, Cs-135, and perhaps Se-79) requires that the waste salt be converted to a form that limits radioactive releases for thousands of years.

Work is underway to develop techniques for consolidating the salt-zeolite mixture into a suitable waste form. Encapsulation of chlorides in an aluminosilicate matrix appears to be a promising technique, which has the advantages of being compatible with pyroprocesses and not requiring steps for converting chlorides into other compounds. An especially promising method for immobilizing salt is to convert the salt-zeolite mixture into a sodalite cage so that the fission product cations become part of the sodalite molecule. Irradiation of salt-occluded zeolites by a Co-60 source at total doses up to 100 Mrad did not measurably change their properties.[10]

METAL WASTE TREATMENT AND IMMOBILIZATION

Four metal waste streams are treated to separate cadmium for return to the process and to produce a metal waste form: (1) the cadmium solution from salt stripping, which contains the rare earths and some uranium; (2) a portion of the cadmium pool in the electrorefiner, which contains some of the noble metals soluble in cadmium (zirconium, palladium, silver, indium, tin, antimony, and tellurium); (3) a steel filter element, which contains insoluble noble metals and the impurities removed from the electrorefiner; and (4) cladding hulls, which also contain some noble metals and zirconium. The hulls have been washed with cadmium to remove adhering actinide-laden electrolyte. In the current concept for treating these streams, they are combined in a retort, and essentially all of the cadmium and salt are vaporized. A matrix metal, such as copper or a copper alloy, may be added during retorting to aid in dissipating fission product heat and to begin the process of forming a metal waste form. After retorting has been completed, additional matrix alloys may be added to disperse and immobilize the wastes. A dense metal ingot is formed by melting or hot pressing.

Initial investigations suggest that metal-alloy waste forms have considerable promise for immobilizing noble metal and rare earth fission products and small amounts of actinides in the combined metal wastes. The metal wastes are made less susceptible to reaction with repository groundwater by dissolving them in a corrosion-resistant matrix alloy that forms stable intermetallic compounds with the fission products. Possible matrix alloys include copper, aluminum, or stainless steel. The steel would be obtained from the fuel cladding and reactor assembly. It seems advantageous to contain technetium as a metallic alloy, because the reducing conditions surrounding a metal waste form should slow the formation of the water-soluble pertechnetate anion. However, the waste form is likely to be highly heterogeneous. Even if the alloy is melted, a large number of intermetallic compounds and insoluble metal phases will precipitate as this highly complex mixture is cooled. The grain size and degree of phase separation will depend, to a large extent, on the cooling rate. The waste form can be encapsulated in a thick-walled container of a corrosion-resistant material to provide nearly complete containment for several hundred years, but long-term resistance to releases into the near-field environment must be provided by the waste form itself. Work has been started to determine fission product release rates from candidate metal alloys.

DEVELOPMENT STATUS AND PLANS

The Fuel Cycle Facility (FCF) of the Experimental Breeder Reactor-II (EBR-II) in Idaho is being readied for a large-scale demonstration of a prototype IFR fuel cycle. The primary objective of this demonstration, which should start in ·1993, is to obtain information for the design of a commercial IFR power plant with a closed fuel cycle and an estimate of its cost. The equipment to disassemble spent EBR-II fuel assemblies, chop fuel elements for feeding to the electrorefiner, consolidate electrorefiner products, and manufacture new fuel assemblies from the recovered actinides have been built and are being installed in the FCF. The chemical and engineering feasibilities of the electrorefining process have been established in tests with unirradiated materials.[11] An electrorefiner capable of handling 20-kg batches of spent EBR-II fuels has been built and should be operational in the FCF by the end of 1993. It is planned that the equipment to demonstrate treatment and packaging of the high-level process wastes will be in operation by 1995, when sufficient volumes of spent salt and metal have been accumulated.

The basic flowsheet for treating and packaging process wastes has been designed, as described in this paper, and laboratory development of several important steps is ongoing. The chemical feasibility of spent salt treatment has been established, and engineering-scale prototypes of the pyrocontactor and salt stripper will be in operation soon with uranium and nonradioactive fission products.

The extraction equipment being developed for liquid salt and cadmium is similar to a centrifugal contactor used for aqueous/organic extractions.[13] The pyrocontactor consists of a hollow rotor spinning at speeds up to 2000 rpm in a stationary housing. Liquid salt and cadmium are fed into the annulus, where they are stirred vigorously and are then drawn into the bottom of the rotor, which acts as both a pump and phase separator. The two liquids are separated by underflow-overflow weirs at the top of the rotor, leave the contactor, and flow to the adjacent contactors in the countercurrent flow train.

A salt stripping system, which is a prototype of equipment for the FCF, is being operated.[9] About 90 kg of salt, which contains about 2 wt % U, Ce, Nd, and Y, was pumped from a large-scale electrorefiner into the stripper. The stripper is a well-stirred vessel, 40 cm in diameter, with several ports for handling instruments, extracting salt or metal samples, and transferring salt or cadmium. The rate and extent of uranium and rare earth reductions are being measured. Techniques are being developed to avoid the formation of difficult-to-dissolve intermetallic compounds by controlling the lithium metal addition rate to the cadmium. In addition, the capability of a sintered metal filter for removing insoluble materials from the salt and metal in the stripper and electrorefiner will be evaluated.

Development of suitable salt and metal waste forms is at an early stage. Laboratory experiments have shown that adequate fission product removals from salt can be achieved with zeolites, but quantitative data are needed to select the best zeolite and to design the process. Techniques must be developed to consolidate salt-zeolite mixtures into leach-resistant, monolithic waste forms. Efforts are beginning to select suitable metal waste forms and develop the fabrication processes.

The arduous work of qualifying salt and metal waste forms for disposal lies ahead. An important objective of the IFR Program is to develop a data base on waste form performance that will be the foundation of the regulatory qualification process for wastes from a future commercial IFR. Prototype forms will be made from the FCF wastes and tested to obtain the data needed for predictions of repository performance. Qualification of IFR wastes may be somewhat less arduous than the qualification of vitrified defense waste, because of their lower TRU content and the ongoing work to qualify high-level wastes planned for disposal at Yucca Mountain.

The ultimate aim of the IFR Program is to lay the groundwork for the construction of future IFRs. A key part of this demonstration program is the development of practical waste treatment techniques and acceptable waste forms.

REFERENCES

1. C. E. Till and Y. I. Chang, "Evolution of the Liquid Metal Reactor: The Integral Fast Reactor (IFR) Concept," in Proceedings of the American Power Conference, Chicago, IL, April 1989, Vol. 51, pp 688-691.

2. C. E. Till, W. H. Arnold, and J. D. Griffith, "The U.S. Liquid Metal Reactor Development Program," presented at American Nuclear Society Topical Meeting -- Safety of Next Generation of Power Reactors, 1988 (CONF-880506-8).

3. J. P. Ackerman, "Chemical Basis for Pyrochemical Reprocessing of Nuclear Fuel," I&EC Research 30(1), 141 (1991).

4. T. R. Johnson and J. E. Battles, "Waste Management in IFR Fuel Cycle," Proceedings of Waste Management '91, Vol. 1, p 815, Arizona Board of Regents (1991).

5. T. Koyama, T. R. Johnson, and D. F. Fischer, "Distribution of Actinides in Molten Chloride Salt/Cadmium Metal Systems," J. of Alloys and Compounds, 189(1), 37-44 (October 1992).

6. J. P. Ackerman and J. L. Settle, "Partition of Lanthanum and Neodymium Metals and Chloride Salts between Molten Cadmium and Molten LiCl-KCl Eutectic," J. of Alloys and Compounds 177, 129-141 (1991).

7. J. P. Ackerman and J. L. Settle, "Distribution of Plutonium, Americium, and Several Rare Earth Fission Products Elements between Liquid Cadmium and LiCl-KCl Eutectic," J. of Alloys and Compounds, in press.

8. J. P. Ackerman, "PYRO - A System for Modeling Fuel Reprocessing," presented at American Nuclear Society Winter Meeting, San Francisco, CA, November 26-30, 1989.

9. E. L. Carls, R. J. Blaskovitz, T. Ogata, and T. R. Johnson, "Tests of Prototype Salt Stripper System for IFR Fuel Cycle," GLOBAL '93, American Nuclear Society International Topical Meeting - Future Nuclear Systems: Emerging Fuel Cycles and Waste Disposal Options, Seattle, WA, September 1993.

10. M. A. Lewis and L. J. Smith, Argonne National Laboratory, personal communication, 1992.

11. Z. Tomczuk, J. P. Ackerman, R. D. Wolson, and W. E. Miller, "Uranium Transport to Solid Electrodes in Pyrochemical Reprocessing of Nuclear Fuel," J. of Electrochem. Society 139 (2), 3523 (1992).

12. E. C. Gay and W. E. Miller, "Plant-Scale Anodic Dissolution of Unirradiated IFR Fuel Pins," GLOBAL '93, American Nuclear Society International Topical Meeting - Future Nuclear Systems: Emerging Fuel Cycles and Waste Disposal Options, Seattle, WA, September 1993.

13. R. A. Leonard, "Recent Advances in Centrifugal Contactor Design," Separation Science and Technology 23, 1473-1487 (1988).

SOIL*EX[SM] - AN INNOVATIVE PROCESS FOR TREATMENT
OF HAZARDOUS AND RADIOACTIVE MIXED WASTE

Gregory C. Gilles[1], Matt Husain[1], Robert Hemmings and Ramona Neuman[1]

RUST Remedial Services Inc.
Clemson Technical Center
Clemson Research Park
100 Technology Drive
Anderson, SC   29625

INTRODUCTION

An unprecedented challenge exists within the U.S. Departments of Energy and Defense to remediate large volumes of soil and debris contaminated with radioactive and hazardous constituents. To meet this challenge, Chemical Waste Management (CWM) has developed the SOIL*EX[SM] (named after SOIL EXtraction) process, an innovative mixed waste treatment scheme for separating and removing radionuclides and metals while destroying volatile organic compounds often associated with soil and debris.

The patented SOIL*EX[SM] treatment process incorporates constituent-specific aqueous-based chemical extraction with solids separation, evaporation, and catalytic oxidation of organics in the vapor phase.   The configuration discussed utilizes the selective ACT*DE*CON[SM] (named after ACTinide DECONtamination) solution and surfactant chemistry in the extraction scheme in conjunction with CWM's demonstrated PO*WW*ER[TM] (the name of an evaporation/catalytic oxidation process) technology for concentration and volume reduction of the extraction system blowdown. The process has some distinct advantages over other alternatives such as conventional soil washing, solvent extraction, or strong acid dissolution.

Physical soil washing processes can be quite effective in reducing the volume of contaminated material from certain types of contaminated soils. These processes are based primarily on physical wet classification of the coarse material (sand) from the fine fraction (clay and silt) that contains the majority of adsorbed contaminants. It is well documented that the majority of organic and inorganic contaminants concentrate in the fine fraction of soils having the highest total surface area.

The success of traditional soil washing processes is based on the amount of coarse material that is generally easier to separate, clean, and hence, release as "clean material." Soil washing processes are generally impractical, and not cost effective for soils with a

---

[1]Gregory C. Gilles, Matt Husain, and Ramona Neuman are no longer employed with Rust Remedial Services Inc.

Chemical Pretreatment of Nuclear Waste for Disposal, Edited by
W.W. Schulz and E.P. Horwitz, Plenum Press, New York, 1995

145

high clay and silt content (40% minus 65 micron) since the fine fraction and concentrated liquid wash water must undergo secondary treatment.

Strong acid dissolution can be utilized as a primary or secondary (fines) treatment alternative. Commonly with radioactive species, concentrated nitric or sulfuric acid is utilized to leach contaminants from the soil. Although strong acids can remove actinides from soil and wastes, strong acids are not selective, and, therefore, will also remove large quantities of minerals containing iron, aluminum, calcium, and magnesium. Strong acid treatment removes large quantities of non-hazardous soil components in the process of removing the radionuclides.

The SOIL*EX[SM] process has the following features:

- A uniquely integrated chemical/physical scheme using commercially available equipment
- Highly selective ACT*DE*CON[SM] dissolution chemistry for removing radioactive constituents from the waste matrix
- Surfactant enhanced chemistry for removal of organics and toxic metals
- Commercially demonstrated PO*WW*ER[TM] technology for concentration of radioactive and inorganic contaminants and destruction of volatile organics.

The process not only removes the alpha emitting actinides, but also separates other hazardous contaminants of concern including volatile organics (VOCs), toxic metals, and nitrate with maximum volume reduction. Figure 1 is a simplified process block diagram of the integrated treatment system.

The SOIL*EX[SM] waste treatment system consists of the following major components or subsystems:

- EXTRACTION to chemically separate the contaminants from the matrix using proprietary dilute aqueous-based chelants and other additives
- DEWATERING for solid/liquid separation of the treated material
- PO*WW*ER[TM] technology utilizing evaporation for concentration of radionuclides, inorganic salts, and heavy metals, and catalytic destruction for volatile organic compounds

The Extraction Subsystem moves the sorbed radioactive and hazardous contaminants from the solid to the aqueous phase by incorporating waste specific selective chemistry. The contaminants in the process are accumulated in the first stage extractor, thereby reducing contaminant-laden blowdown to the PO*WW*ER[TM] Subsystem. The extraction scheme differs from traditional soil washing since volume reduction of the waste stream is not limited by the fraction of fines in the waste. The extraction hardware and ACT*DE*CON[SM] reagents facilitate true chemical extraction that removes contaminants from both the coarse and fine components providing enhanced capability for significant volume reduction without secondary waste treatment.

The second component of the proposed treatment process is the Dewatering Subsystem composed of a conditioning step, dewatering units, and a clarification system. This system separates liquids from treated solids delivering a high solids treated material. An important part of this dewatering phase is the step in which the solids are washed with high quality water from the PO*WW*ER[TM] Subsystem. This step adds a higher degree of confidence that the final treated solids will meet the prescribed treatment standards.

The last major component of the proposed treatment system is the PO*WW*ER[TM] Subsystem. This PO*WW*ER[TM] Subsystem, a demonstrated and proven treatment technology, provides treatment and volume reduction of contaminant-laden blowdown from the extraction system. The main components comprising this subsystem include a forced circulation evaporator, a catalytic oxidizer using a proprietary catalyst, and a steam condenser.

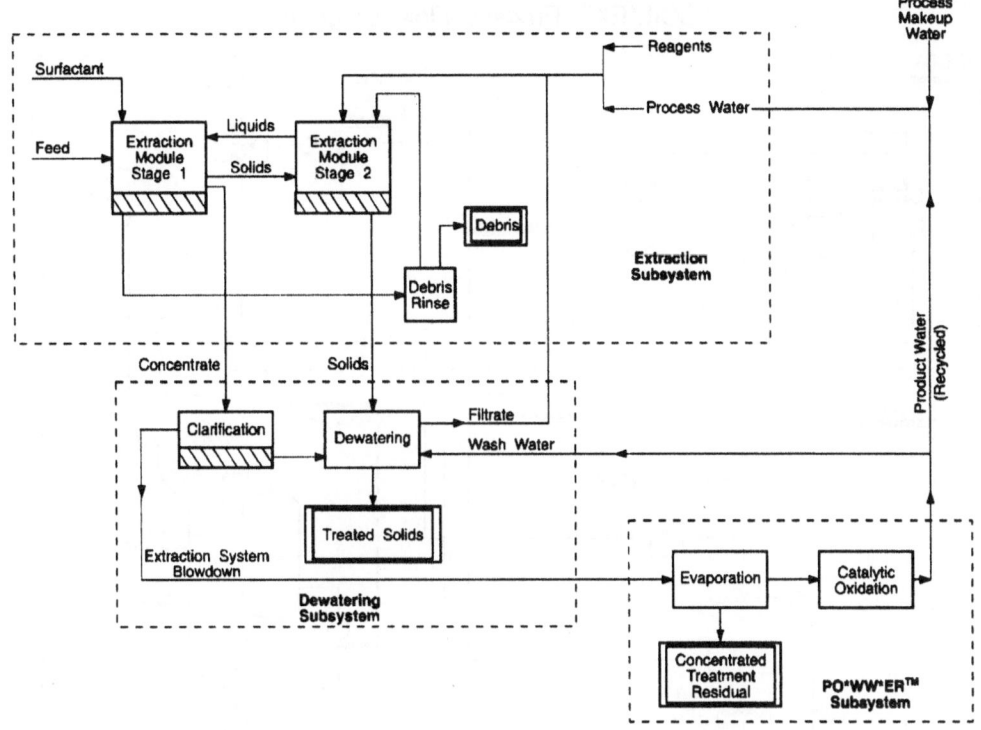

Figure 1. SOIL*EX^{SM} Block Diagram Integrated Treatment System

Overall, the primary benefits of the SOIL*EX^{SM} process include:

- well-tested process chemistry for dissolution of plutonium, americium, other radionuclides, as well as removal of organic and other hazardous contaminants from wastes
- uniquely combined commercially available equipment and systems
- flexibility to accommodate fluctuations in the incoming contaminant levels
- maximum volume reduction exceeding 95%
- complete recycle of process wastewater
- no secondary waste generation requiring further treatment or disposal
- minimal use of extractant chemicals
- versatility to treat a wide array of contaminants in soil and debris including organics and toxic heavy metals

EXTRACTION

The main function of the extraction subsystem is to promote mass transfer of the contaminants (i.e., actinides, VOCs, toxic metals) into the aqueous phase. The extraction process chemistry incorporates the proven ACT*DE*CON^{SM} solution for removal of actinides and nonionic surfactant addition for dissolution of organics. The selected waste specific chemistry is introduced within the framework of an innovative countercurrent solid/liquid extraction scheme.

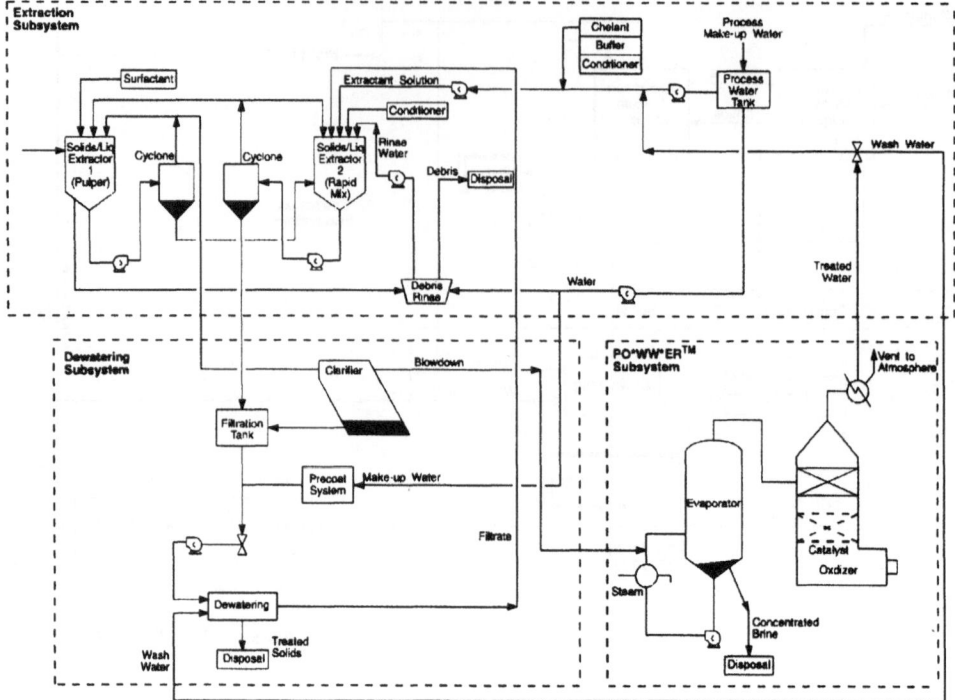

Figure 2. A process schematic of integrated treatment system showing the major components and functional subsystems.

The two-stage extraction subsystem will (1) provide excellent contact of the contaminated solids with the ACT*DE*CON™ reagents and surfactant; (2) accumulate contaminants in the first stage thus reducing blowdown to the PO*WW*ER™ subsystem; (3) contact the solids after being treated in the first stage with fresh chemicals for further dissolution in the second stage; and (4) minimize the use of chemicals and make-up water.

Feed to the process can consist of contaminated soils, metallic debris, plastic, inorganic salts, as well as some cellulosic and combustible material. The feed is subjected to a two-stage countercurrent extraction configuration that will remove the radionuclides, volatile organics, and toxic heavy metals into solution. Solid/liquid separation is aided by the use of hydrocyclones to generate a high solids underflow stream that will be sent for dewatering. A small overhead blowdown stream containing the concentrated contaminants is sent to the PO*WW*ER™ Subsystem for concentration and volume reduction.

An enclosed screw conveyor transports the material into the first solid/liquid extractor. The first stage extractor serves two main purposes. Inorganic debris passing a 5 cm (or 2") screen requires some further sizing and mixing. The first extractor is a pulper-type that is equipped with a hardened shear-type agitator (rotor) capable of breaking up and sizing hard solids and debris to the smaller size required. Intimate mixing is also a desirable feature of the system since this improves the rate and efficiency of mass transfer of contaminants to the aqueous phase. Most of the contaminant dissolution occurs in the first stage. Extraction residence time is relatively short. An inert nitrogen atmosphere is maintained in the head space of the first extractor to eliminate any potential fire hazard when combustible organic compounds are present in the feed.

The pulper-type mixer is preferred for most applications because it not only reduces the size of materials in the feed, but provides mechanical, hydraulic, and attrition energy to the solids/liquid mixture for extraction. The objective of the pulper is to process the materials in a slurry. As the solids are continuously conveyed into the extractor, chemical components are introduced to drive the extraction of contaminants. Chemical components include a nonionic surfactant for enhancing organic contaminant removal and the ACT*DE*CON^{SM} reagents. Depending on the alkalinity and radioactive constituents of concern, the ACT*DE*CON^{SM} solution is derived from the following: chelating agents such as dilute EDTA (Ethylene Diamine Tetra Acetic Acid) or citrate; a carbonate buffer; and a conditioner such as air or hydrogen peroxide. These reagents along with makeup water and the contaminated solids comprise a low concentration slurry in the first extractor.

The rotor, which fits into the bottom of the extractor, has a perforated screen on the bottom to filter oversize particles exiting the extractor. These larger solids travel counter-currently through both stages of extraction and through the hydrocylones before exiting the system. Following the time in the first extractor, the slurry is transferred to the second stage conventional baffled mix tank to allow longer and less vigorous solids/liquid contact time.

The first extractor is also equipped with a specially designed debris trap. This automated debris collector is used to remove inorganic debris, rocks, plastic, or cellulosic materials that are not sized by the rotor. This oversized debris from the extractor collects in the trap for periodic removal.

Solid debris removed from the trap is subsequently collected and rinsed in a designated vessel using recycled (treated) process water. The debris trap is emptied periodically as the capacity is reached during operation. Low pressure water spray nozzles are located inside the screened collection container to decontaminate the debris fraction with the rinse water. The rinse water used to decontaminate the debris is sent to the second stage extraction tank. Solid debris is then tested to determine the final disposal options.

DEWATERING

The dewatering system is essentially comprised of recessed chamber plate and frame pressure filtration or centrifugation and clarification equipment to separate the treated solids. The dewatering system also incorporates a solids washing step with high quality water before the solids are discharged. This ensures a higher degree of total extraction efficiency and minimizes the loss of extractant chemicals with the retained liquid with the solids.

The underflow exiting the hydrocyclone separators carries the treated solids from the extraction stage in approximately a relatively high solid content slurry. The slurry is fed to a filtration conditioning tank. Since the extraction operation is continuous, this tank serves both as a surge tank and as a chemical conditioning tank. A mechanical mixer is provided for gentle agitation and suspension of the slurry. The suspension ensures a uniform feed to the dewatering equipment.

A concentrated blowdown stream containing some suspended fine solids is taken from the first cyclone. Chemical pretreatment, such as a polyelectrolyte addition to condition the solids for clarification and subsequent separation, may be required. Any sludge from the clarifier is combined with the solids underflow stream in the filtration conditioning tank for dewatering.

Table 1. Capabilities of PO*WW*ER™

Specific Capabilities of PO*WW*ER™	Demonstrated Results	Comparison to Conventional Technology
High volume reduction of waste materials.	Concentrated wastewaters from 0.5% to 65% total solids which provides a volume reduction of 130 to 1.	Reverse Osmosis: Limited to 7-8% total solids and requires substantial pretreatment. Volume reductions of 40 to 1 would be typical.
High removal efficiency for organic pollutants.	Routinely removes 99% of all organic compounds. Higher removals are achieved for specific pollutants (e.g. 99.99% for toluene).	Biological Treatment: Limited to 95-98% for biologically degradable compounds; others are not removed.
High removal efficiency for heavy metals.	Heavy metals are separated from the effluent by evaporation; therefore, PO*WW*ER™ can achieve extremely high removal efficiencies.	Precipitation: Limited to the solubility point of metallic species.
Significant flexibility to handle changes in feed composition.	System is insensitive to changes in heavy metal concentration and only sensitive to major changes in organic composition.	General: Most systems cannot accommodate changes in feed composition while assuring high quality effluent.
Minimal pretreatment is required.	Only possible pretreatment may be neutralization and control of foaming tendencies.	General: Comprehensive pretreatment may be required for sophisticated treatment processes.

Clean process wash water (free of hazardous and radioactive constituents) is used for a final wash of the solids prior to discharge. This added feature allows the solids to make contact with clean water one final time before the solids are discharged. The final washed solids exiting the subsystem are approximately 50 - 75% by weight solids. The treated solids can then be sampled for specific radionuclides and hazardous constituents and any other appropriate parameters to determine whether the treatment goals have been achieved.

PO*WW*ER™

A central feature of the proposed treatment system is the PO*WW*ER™ Subsystem for treatment and volume reduction of the contaminant-laden blowdown from the extraction system. The capabilities of the PO*WW*ER™ technology are listed in Table 1. This subsystem consists of a forced circulation evaporator, a catalytic oxidizer using a proprietary catalyst, and a steam condenser.

The blowdown from the extraction subsystem is fed into the evaporator where it is concentrated, thus reducing the volume of contaminant residue by nearly 100 to 1. Water and volatile organic and inorganic compounds vaporized in the evaporator are directed into the oxidizer. The volatile compounds including chlorinated compounds, aromatics, sulfides, ammonia, cyanides, etc. are completely oxidized. Acid gases such as Hydrochloric Acid (HC1) may be formed during the oxidation of chlorinated organics. In these applications, a conventional scrubber is added between the oxidizer and condenser, and scrubber blowdown is returned to the evaporator. The treated vapors are then condensed in the condenser producing a high quality product water. The product water is recycled into the extractor and/or utilized in the dewatering process for washing. The concentrated contaminants in the form of a small brine volume are delivered from the evaporator for final disposal. Volume reduction can be further enhanced by drying of the brine in the collection containers.

DEVELOPMENT STATUS

The SOIL*EX℠ process uniquely combines surface enhanced chemistry for organics removal, a highly selective dissolution chemistry for radionuclide removal, and a concentration/volume reduction technology for treatment of residues to provide a state-of-the-art approach for treatment of hazardous and radioactive mixed waste. The key elements of the integrated SOIL*EX℠ process are:
- ACT*DE*CON℠ chemistry
- PO*WW*ER™ technology

ACT*DE*CON℠ is a dilute aqueous extraction chemistry which has been developed from the extensive work done in the United Kingdom for removal of actinides from Magnox reactor fuel cladding from the mid 1970's to 1989 involving British Nuclear Fuels (BNFL), Central Electricity Generating Board (CEGB), South of Scotland (SSEB), and the UK Atomic Energy Agency (UK AEA). It was adapted for soil applications and successfully demonstrated by Bradtec for the U.S. Department of Energy (DOE) in December, 1991. RUST Remedial Services Inc. has exclusive license rights to the ACT*DE*CON℠ technology.

Preliminary bench scale tests performed on soil from a DOE National Laboratory site spiked with plutonium, americium and selected organics using ACT*DE*CON℠ reagents and nonionic surfactant have shown satisfactory results (Table 2). Note that the decontamination factor (DF) is the ratio of the starting concentration of the contaminant to the final concentration of that contaminant after treatment, and is usually applied to radioisotope applications.

Table 2. Preliminary Bench Scale Test Results ACT*DE*CON^SM Chemistry.

Contaminant	Initial Soil (Spiked)	Treated Soil	Decontamination Factor (DF)
Pu-238	94.75 Bq	8.75 Bq	10.8
Am-241	61.03 Bq	1.99 Bq	30.7
TCE	59 ppm	4.5 ppm	92.4*
PCE	210 ppm	22 ppm	89.5*
CCl₄	27 ppm	3 ppm	88.9*
TPH	2000 ppm	335 ppm	95.2*

*Denotes % removal

TCE - Trichloroethylene   Pu - Plutonium   Am - Americium
PCE - Perchloroethylene   CCl₄ - Carbontetrachloride   TPH - Total Petroleum Hydrocarbons

The PO*WW*ER™ technology has been a proprietary development of Chemical Waste Management (CWM) starting with initial conceptualization in 1987. The technology has been well demonstrated through the operation of a fully integrated pilot plant at CWM's facility in Lake Charles, Louisiana. It is now included in the U.S. Environmental Protection Agency's (EPA) Superfund Innovative Technologies (SITE) program for 1992. A list of PO*WW*ER™ plants built to date is as follows:

PO*WW*ER™ Installations

Location	Size	Plant Start-up	Commercial Use
Lake Charles, LA	0.5 gpm	February 1988	Demonstration
CTC, Anderson, SC	1.5 gph	December 1991	Demonstration
Ysing Yi Island, Hong Kong	50 gpm	December 1992	Commercially

In addition, feasibility studies and designs are underway for more than a dozen PO*WW*ER™ facilities around the world ranging in capacity from 10 gpm to 300 gpm. Tables 3 and 4 summarize the results of PO*WW*ER™ pilot plant testing on complex landfill leachates containing a wide variety of organic and inorganic contaminants. The data shows in excess of 99% of destruction of organics and an average of 200 to 1 reduction in the metals.

Table 3. Sample results from PO*WW*ER™ Pilot Plant Testing on Leachate

	Leachate A		Leachate B	
	Feed	Product Water	Feed	Product Water
TOC (ppm)	1,350	15	500	3
TDS (ppm)	15,000	40	6,000	76
Total Organic Pollutants[1] (ppm)	140	0.1	2.6	<0.1

[1]Organic Pollutants: Methylene Chloride, 1,1,1-Trichloromethane, Benzene, Toluene, Acetone, Tetrachloroethylene, 4-Methyl-2-Pentanone, 2-Butanone, 1,2-Dichloroethane, 1,1,2-Trichloromethane
TOC - Total Organic Carbon   TDS - Total Dissolved Solids

Table 4. Sample results from PO*WW*ER™ Pilot Plant testing on Leachate

Metal	Brine	Product Water (PPM)
Arsenic	300 - 750	<0.09
Calcium	1,100 - 3,600	<11
Copper	7 - 40	<0.03
Magnesium	160 - 370	<2.3
Potassium	80,000 - 160,000	<3.8
Sodium	60,000 - 90,000	<19

After a comprehensive evaluation of alternate technologies for the Idaho National Engineering Laboratories' (INEL) PIT 9 project for the treatment of soil and buried mixed waste, the DOE has selected the SOIL*EX™ process for proof-of-process (POP) testing. A fully integrated SOIL*EX™ pilot plant has been assembled at RRS's Clemson Technical Center in Anderson, SC. The pilot plant was completed by the mid-1993. POP testing was performed in the last half of 1993 using soils and sludges treated to simulate actual PIT 9 soils and sludges. This testing will confirm the applicability of the SOIL*EX™ process and provide engineering data to design and build a full scale plant for the treatment of INEL PIT 9 soils and sludges. During the POP testing, various configurations of the mixing and contacting equipment were identified, and an optimum arrangement was determined to be somewhat different from that described above. However, the basic operations remained unchanged.

APPLICABILITY

The SOIL*EX[SM] technology is ideally suited to handle wastes containing a mixture of the contaminants and pollutants identified in Table 5.

Table 5. Contaminants and Pollutants Suitable For SOIL*EX[SM] Treatment

Organic	Inorganic	Radioactive
• Halogenated volatiles	• Heavy metals	• Plutonium
• Halogenated semi-volatiles	• Non-metallic toxic elements	• Americium
• Non-halogenated volatiles	• Cyanides	• Uranium
• Non-halogenated semi-volatiles	• Ammonia	• Technetium
• Organic pesticides	• Nitrates	• Thorium
• Solvents	• Salts	• Radium
• (Benzene Toluene Thyl Benzene Xylene (BTEX)		• Barium
• Organic cyanides		
• Non-volatile organics		

Some examples of generally recognizable hazardous and mixed wastes at various Department of Energy facilities/laboratories which can be effectively treated by the SOIL*EX[SM] process are shown in Table 6.

Table 6. DOE Facilities With Wastes Suitable For SOIL*EX[SM] Treatment

DOE Facility/Laboratory	Examples for SOIL*EX[SM] Applications
Argonne	570 holding pond, CP-5 reactor
Ames	Chemical disposal site
Brookhaven	Former landfill soils
Oak Ridge	K-25 area soils, X-10 soils
Weldon Springs	Raffinate pit sludge
Fernald	Operable Unit #3
Savannah River	Miscellaneous basins
Idaho	PIT 9, TSF disposal pond
Richland	100, 200, 300 area soils and wastes
Los Alamos	Canyons - sediments, drainfields
Sandia	Chemical and mixed waste landfills
Nevada	Pu contaminated soils
Rocky Flats	903 pad, solar ponds, off-site soils

CLEAN OPTION: AN ALTERNATIVE STRATEGY FOR HANFORD TANK WASTE REMEDIATION; DETAILED DESCRIPTION OF FIRST EXAMPLE FLOWSHEET

J. L. Swanson

Pacific Northwest Laboratory
Richland, Washington 99352

INTRODUCTION

Disposal of high-level tank wastes at the Hanford Site is currently envisioned* to divide the waste between two principal waste forms: glass for the high-level waste (HLW) and grout for the low-level waste (LLW). Several approaches have been proposed to accomplish this division; these approaches lead to a range of volumes of both the HLW and LLW fractions and a range of long-term risks associated with the disposed LLW. These approaches do not include emphasis on the waste minimization issues that are involved in the "clean option" discussed in this paper.

The waste processing system described by Grygiel et al. (1991) will be used as a "reference case" for comparison with the "clean option" presented here. In the reference case, chemical separations have been proposed for the tank wastes that are sufficient to prevent the grout from becoming a transuranic (TRU) waste form according to U.S. Nuclear Regulatory Commission (NRC) Class C LLW limits (10 CFR 61). However, in this case, tens of megacuries of other radioactive isotopes would become part of the grout. In addition to total radioactivity, some mobile and long-lived radionuclides and the high nitrate content contribute to the long-term risks associated with the grout disposal. For the HLW fraction, nonradioactive waste constituents would lead to HLW glass volumes of over 10,000 canisters. These canisters would cost several billion dollars to produce and dispose of in a geologic repository. When combined with canisters expected from other sites, the total volume would seriously impact the space presently planned for DOE defense wastes at the Nevada repository.

This paper describes the results of a study led by Pacific Northwest Laboratory** to determine if a more aggressive separations scheme could be devised which could

---

*After this paper was written, the vision for the LLW form was changed from grout to glass. This change has little effect on the major points of this study.

**Pacific Northwest Laboratory is operated for the U.S. Department of Energy by Battelle Memorial Institute under Contract DE-AC06-76RLO 1830.

Chemical Pretreatment of Nuclear Waste for Disposal, Edited by
W.W. Schulz and E.P. Horwitz, Plenum Press, New York, 1995

mitigate concerns over the quantity of the HLW and the toxicity of the LLW produced by the reference system. This more aggressive scheme would meet NRC Class A LLW limits (10 CFR 61) and would decrease the volume of HLW glass by a factor of at least 10. Additional benefits might result from using HLW and LLW disposal forms other than glass and grout, but such departures from the reference case are not included at this time.

The evaluation of this aggressive separations scheme addressed institutional issues such as:

- radioactivity remaining in the Hanford Site LLW grout
- volume of HLW glass that must be shipped offsite
- disposition of appropriate waste constituents to nonwaste forms
- generation and disposition of secondary wastes
- emissions.

Specific goals for an aggressive separations scheme were developed based on these issues. These goals comprise the clean option. Once the specific goals were defined, a chemical process flowsheet for a separations scheme was developed as an example of how the clean option could be achieved. The flowsheet was used as a basis to assess waste volumes, to consider the feasibility of the process chemistry, and to identify technical issues that must be resolved to achieve the clean option goals. Although the specific example flowsheet was used to guide the identification of technology issues, alternative flowsheet and technology approaches were considered sufficiently to determine that the issues would be the same for a range of approaches.

The first clean option example flowsheet presented here is based on the concurrent processing of all portions of the tank wastes. Because of the currently higher degree of technical uncertainties in the sludge processing area, it could be beneficial to begin treating the supernate and salt cake portions of the waste while deferring sludge processing for several years to allow time for the uncertainties to be resolved. Such a "decoupled" or "phased implementation" approach would require some flowsheet modifications.

Other flowsheet modifications could be required if the waste processing were to be done in a number of small, field-deployed units rather than in a large central facility. Process simplification, even if at the expense of process performance, could be very important to such modifications.

The details presented in this paper are a first step in providing the information needed to define the processing operations in sufficient detail to allow a facility cost estimate to be made. This paper presents the stage of development as of December 1992. Subsequent efforts have led to some flowsheet modifications (generally simplifications), but the basic features are the same. Discussion of those modifications will await completion of those efforts.

Finally, it should be reemphasized that the clean option is not a flowsheet but, rather, is a set of goals. The flowsheet details presented here are for the first example flowsheet, based on the initial assumptions of the study. The use of other assumptions regarding the nature of the HLW and LLW disposal forms could give significant improvements in both flowsheet simplicity and in predicted long-term risk. Assumption of a third type of waste, a "decay storage" waste for Sr-90 and Cs-137 (and perhaps Ni-63), would also allow significant flowsheet simplification. Incorporation of such potential improvements is left to await the outcome of additional study; in this paper we continue to assume grout to be the LLW form and borosilicate glass of the composition currently planned for the Hanford Waste Vitrification Plant (HWVP) to be the HLW form.

STUDY BASES

The primary goal of the effort discussed in this paper was to develop a reasonable example flowsheet describing some process steps typical of those likely to be required to meet the objectives of the clean option. This flowsheet could then be used in preliminary evaluations of technical feasibility, impacts on waste volumes, and costs. This section describes the objectives and goals to be met and the wastes to be treated, and then describes the separations process requirements that are necessary to process the waste in accordance with these goals and objectives.

Clean Option Objectives

The Tank Waste Remediation System planned for Hanford encompasses several options, including the clean option. The broad objectives of the clean option compared with the other options being considered are listed below:
- minimize the residual radioactivity and chemical toxicity of the bulk of the waste, most often referred to as LLW, to further reduce the long-term environmental effects on the Pacific Northwest region
- decrease the volume of the HLW to reduce long-term demands on the geologic repository space
- accomplish the first two goals without increasing the amount of waste requiring disposal.

The specific goals developed to achieve these objectives are given in the following section.

Specific Goals for First Example Clean Option Flowsheet. The following goals have been adopted as the basis for formulating an aggressive but feasible strategy for separating the components of the waste to meet the clean option objectives:
- The radioactivity in the waste will be removed from the bulk of the waste such that the radionuclides in the remaining LLW will not exceed NRC Class A maximum allowable concentrations for shallow land burial of radioactive materials.
- Even though not required in order to meet the Class A limits, the concentrations of technetium, iodine, and uranium are reduced on an as low as reasonably achievable (ALARA) basis.
- Additional radioactivity will be removed from the LLW where significant reductions can be achieved through minor modifications to the process scheme.
- The LLW will be disposed of in a manner that complies with U.S. Environmental Protection Agency (EPA) and state regulations for disposal of hazardous wastes.
- The radionuclides that have been removed from the bulk of the waste will be disposed of within about 1000 canisters of a borosilicate glass that meets current HWVP glass specifications.
- Waste minimization principles will be used to minimize the volume of LLW.

Assumptions and Rationale for Selection of Goals. One of the principal objectives of the clean option is to reduce the radioactivity and toxicity in the bulk of the waste to a level that would be publicly acceptable and would reduce the long-term environmental effects on the Pacific Northwest region. Drinking water standards were initially considered as the limit for the concentration of radionuclides in the LLW stream; however, it was not feasible to achieve the separation factors required to meet these standards. Instead, the NRC Class A specifications were selected as the next most stringent standards that are legally established. In general, these standards represent a high level of risk reduction. However, for the Hanford tank wastes, several assessments (Buck and

Peffers 1991; Droppo et al. 1991) have indicated that, if not retained within the disposal system, Tc-99, I-129, and U-238 are primary contributors to long-term radioactivity-related risk even at concentrations allowed by the Class A limit. Thus, the clean option goals include an additional reduction for these constituents.

Grout is an acceptable waste form for heavy metals and is assumed to be the waste form used for the LLW in the specific clean option example outlined later in the text. To reduce the chemical toxicity from other components, the clean option will include destruction of nitrates, nitrites, and organics prior to grouting. Actual limits on nitrates, nitrites, and organics in the grout are not defined.

Another main objective of the clean option is to reduce the volume of HLW that would require disposal in a geologic repository. The target volume for the HLW, 1000 canisters of glass, appears to be a reasonable one based on the following information:

- 1000 canisters is a quantity large enough to accommodate the heat generation from the radionuclides in the Hanford tank wastes.
- 1000 canisters would yield a significant reduction in demands on repository space.

Borosilicate glass is assumed to be the HLW form. Work has been under way for a number of years in the HWVP project to determine the concentrations of various constituents that can be accommodated by existing borosilicate vitrification technology. Hanford tank waste includes abundant nonradioactive species such as iron, zirconium, and aluminum that would substantially increase the volume of the glass beyond 1000 canisters if not diverted to the LLW stream. Also, less abundant nonradioactive species exist that are sufficiently insoluble in molten glass that their concentrations need to be limited in the feed to the HWVP.

The third principal objective of the clean option is to achieve the first two goals without substantially increasing the amount of waste for disposal. This objective appears achievable through the aggressive use of waste minimization and recovery and recycle of key chemicals used to process the waste, such as nitric acid and sodium hydroxide.

Wastes to be Treated

Hanford tank wastes are generally grouped in two categories, single-shell tank (SST) wastes and double-shell tank (DST) wastes. In actual fact, each of these two groupings contains a variety of wastes that have significant compositional variations. In this study, a feed composition representing the average of all SST wastes was assumed for simplicity in flowsheet development. However, DST wastes were also included in estimating the quantities of waste and the radionuclide contents of the wastes that would result from implementation of the flowsheet.

The total quantities of bulk chemical constituents and important radionuclides present in the SST wastes, as assumed for this study, are given in Table 1. These values were developed in other studies at Hanford and are accepted in this study as a basis for flowsheet development and waste quantity comparisons, with the realization that results of future waste characterization work may give a somewhat different picture. The quantities of bulk components present in the DST wastes are generally smaller than those present in the SST wastes; exceptions are aluminum, fluoride, and zirconium, where comparable quantities are present in the two types of waste. Thus, the data of Table 1 provide a good approximation of the total quantities of bulk components present in Hanford tank waste.

The constituents of the Hanford tank wastes are divided among three categories: supernate, salt cake, and sludge. Supernates are liquid phases that generally contain high concentrations of soluble salts such as sodium nitrate, sodium hydroxide, etc. Salt cakes are solid phases that contain salts that crystallized out as a result of evaporative

Table 1. Bulk constituents of total SST waste.

Component	Amount Present[1]	
	Mole	Gram
Al	9.05×10^7	2.44×10^9
Am	$\sim 4 \times 10^1$	$\sim 10^4$
Ba	$4 \times 10^{6[2]}$	5×10^8
Bi	1.25×10^6	2.61×10^8
Ca	3.25×10^6	1.30×10^8
Cr	2.21×10^6	1.15×10^8
Cs	4.2×10^3	$5.6 \times 10^{5[2]}$
Fe	1.12×10^7	6.27×10^8
Hg	4.49×10^3	9.00×10^5
K	$9 \times 10^{6[2]}$	4×10^8
Lanthanides	$6 \times 10^{5[2,3]}$	8×10^7
Mn	2.18×10^6	1.20×10^8
Na	2.25×10^9	5.17×10^{10}
Ni	3.04×10^6	1.78×10^8
Np	2.0×10^2	$4.7 \times 10^{4[2]}$
Pd	$\sim 2 \times 10^4$	$\sim 2 \times 10^{6[2]}$
Pu	2.4×10^3	$5.8 \times 10^{5[2]}$
Si	7.93×10^6	2.22×10^8
Sr	4.11×10^5	$3.60 \times 10^{7[2]}$
Tc	$\sim 5 \times 10^3$	$\sim 5 \times 10^5$
Th	5.6×10^4	$1.3 \times 10^{7[2]}$
U	5.9×10^6	$1.4 \times 10^{9[2]}$
Zr	2.70×10^6	2.46×10^8
Cl	1.13×10^6	4.00×10^7
CO_3	2.68×10^7	1.61×10^9
F	4.24×10^7	8.05×10^8
$Fe(CN)_6$	1.52×10^6	3.22×10^8
H_2O	2.49×10^9	4.48×10^{10}
I	4.3×10^3	$5.6 \times 10^{5[2]}$
NO_3	1.56×10^9	9.67×10^{10}
NO_2	1.04×10^8	4.80×10^9
OH	5.38×10^8	9.15×10^9
PO_4	9.20×10^7	8.74×10^9
SO_4	1.72×10^7	1.65×10^9

[1]From values in HDW-EIS (DOE 1987) unless indicated otherwise.

[2]Separate estimate.

[3]Midpoint between 1.7×10^6 in HDW-EIS and 2.0×10^5 in separate estimate.

concentration of supernates; they can be dissolved in water. Sludges are solid phases containing primarily insoluble hydroxides that precipitated out when the (acidic) process waste solutions were made basic for storage in the carbon steel underground storage tanks.

The quantities of barium and of the lanthanide elements present in the wastes are of special significance to development of a flowsheet capable of meeting the clean option goals. This is because the chemical similarities between barium and strontium and between lanthanides and americium make their separations difficult and because such separations are required if strontium and americium are to be disposed within ~1000 canisters of HWVP-type glass. As noted in Table 1, the estimated lanthanide value is the midpoint between two estimates that differ by a factor of ~10. The estimated barium content is even more uncertain; some estimates are lower than that given in Table 1 by a factor of several hundred. Recent evaluations by the author indicate that the current "best" estimate should be lower than that given here by a factor of ~10; however, the higher value is retained in this paper to be in accord with Straalsund et al. (1992).

Separations Required to Meet Objectives

Stated in its simplest form, the initial objectives of the clean option would lead to disposal of all radionuclides in two fractions, a LLW fraction that contains very minimal radioactivity, to be disposed of onsite, and a HLW fraction to be disposed of in a geologic repository, with each fraction being environmentally acceptable and having the minimum possible volume. Not surprisingly, the tank contents require that trade-offs be made between radionuclide contents and volumes of the waste fractions. However, the contents do not preclude attainment of the initial goals of having a LLW fraction that meets NRC Class A criteria and a HLW fraction that occupies ~1000 canisters of HWVP-type glass. This is discussed in more detail in the following sections.

HLW-Driven Separations Requirements. The separations requirements that are driven by requirements on the HLW fraction are concerned primarily with the quantities of the bulk chemical constituents of the wastes that can be contained within ~1000 canisters of HWVP-type waste. Estimation of the number of canisters that would result from separate vitrification of each of the waste constituents provides an indication of which constituents can be completely accommodated in ~1000 canisters and, for those that exceed that amount, an indication of the extent of separation that is required from those waste constituents that are to be disposed of in glass.

Table 2 shows estimates of the mass percentage increases in the glass composition allowed by the addition of some individual waste constituents, based on current plans for the HWVP. Many of these estimates were derived from the current HWVP feed specifications, and assume a 25% waste oxide:75% glass frit mixture in the vitrification process. Where values needed for this study were not available from the HWVP information, they were estimated separately as indicated in the table. The actual percentage of sodium in the glass is higher than is indicated here because of the sodium content of the glass frit. It should be possible to increase the sodium content of the waste feed by decreasing the sodium content of the frit, but that is not assumed here.

It should be emphasized that these HWVP-type glass "limits" have not been challenged or evaluated in this study. It is probable that some of them could be increased as a result of additional consideration. Mixed effects that might occur when more than one of these waste constituents is added to the glass at the same time are also not addressed here, except for the intuitive judgment that the total number of canisters required to contain all the waste constituents will be less than the number obtained by summing all the individual values.

Table 3 contains estimates of the number of HWVP-type glass canisters that would be required to dispose of the Hanford SST+DST waste constituents, based on 1650 kg of glass per canister, if none of the constituents were routed to LLW instead. Also listed in Table 3 are estimated percentages of the various waste constituents that could be accommodated in 1000 canisters of HWVP-type glass, again considering each constituent on an individual basis. These data indicate that appreciable separation from the radioactive constituents—TRUs, technetium, cesium, and strontium—will be required

Table 2. Allowable additions to glass in waste feed.

Waste Constituents	Maximum Weight Percent[1]	Basis
Al	3.4	HWVP feed specification
Ba	4.5	HWVP feed specification
Bi	12	Estimate from glass formulation expert
Ca	3.6	HWVP feed specification
Cr	0.34	HWVP feed specification
Cs	23	Molar equivalent to Na, K; heat loading more restrictive
Fe	12	HWVP feed specification
Lanthanides	1.7	HWVP feed specification
Mn	3.2	HWVP feed specification
Na	3.8	HWVP feed specification
Ni	1.6	HWVP feed specification
Pd	0.22	HWVP feed specification
Si	2.0	HWVP feed specification
Sr	4.8	Midpoint of molar equivalence to Ba and Ca
Tc	0.22	Assume the allowed percentage for Pd
Th	7.0	Assume to be analogous to U and Zr
TRU	1.0	HWVP feed specification
U	6.8	HWVP feed specification
Zr	7.4	HWVP feed specification
F	1.7	HWVP feed specification
PO_4	1.3	HWVP feed specification
SO_4	0.7	HWVP feed specification

[1]Gram of constituent per 100 g of glass.

for all of the bulk chemical constituents, except perhaps bismuth, in order to meet the 1000-canister goal for disposal of the radioactive constituents. To be prudent, separations should be performed to give levels even lower than those indicated here; thus, separation of bismuth is also desired.

In addition to the chemical composition limits discussed above, radiolytic heat generation is also a factor in defining the number of canisters needed for disposal of the HLW fraction. Based on a tentative limit of 1500 watts/canister, it would require ~400 canisters to dispose of the Sr-90 and Cs-137 currently contained in the tank wastes. An additional ~300 canisters would be required to dispose of the Sr-90 and Cs-137 that were removed from the waste and are currently stored in capsules. Thus, to obtain a HLW fraction much smaller than 1000 HWVP-type canisters, the Sr-90 and Cs-137 would have to be disposed of in some other fraction.

LLW-Driven Separations Requirements. The concentrations of radionuclides allowed in the three classifications of LLW defined by the NRC are summarized in Table 4. As was discussed earlier, the Class A limits, which are markedly more

Table 3. Effect of important constituents on canister count.

Waste Constituents	Number of Canisters[1]	Allowed Addition to HLW, %[2]
Na	1×10^6	0.1
PO_4	4×10^5	0.2
SO_4	2×10^5	0.5
Cr	8×10^4	1.2
F	6×10^4	1.7
Al	5×10^4	2.0
U	1×10^4	10
Ba	7×10^3	14
Ni	7×10^3	14
Si	7×10^3	14
Lanthanides	5×10^3	20
Zr	5×10^3	20
Fe	4×10^3	25
Ca	2×10^3	50
Mn	2×10^3	50
Pd	1×10^3	100
Bi	1×10^3	100
Sr	5×10^2	100
Tc	5×10^2	100
Th	1×10^2	100
TRU	5×10^1	100
Cs	$<1 \times 10^{1[3]}$	100

[1]Includes both SST and DST wastes.

[2]To meet the goal of 1000 canisters.

[3]Heat loading considerations would require a higher number.

restrictive than the Class B or Class C limits, were selected as the upper concentration level goals for the study, with some additional separations being done on an ALARA basis.

The minimum separation factors that must be achieved to meet the Class A portion of the initial clean option goals can be estimated by 1) estimating the eventual volume of the LLW fraction after the treatment and pretreatment steps, 2) dividing the radionuclide inventories by this volume to obtain the hypothetical concentrations that would result if no separations processing was performed, and 3) dividing these hypothetical concentrations by the Class A LLW limits. Results of such comparisons for the SST wastes and radionuclides having half-lives of >13 years are shown in Table 5. For these comparisons, the volume of the LLW grout was assumed to be governed by the sodium content at a concentration 5 $\underline{M}$.

Table 4. NRC LLW limits for radionuclides present in SST waste.

Radionuclide	Class A	Class B	Class C
	Allowed Concentration, Ci/m^3		
H-3	40	--	--
C-14	0.8	8	8
Co-60	700	--	--
Ni-63	3.5	70	700
Sr-90	0.04	150	7000
Tc-99	0.3	3	3
I-129	0.008	0.08	0.08
Cs-137	1	44	4600
Other[1]	700	--	--
	Allowed Concentration, nCi/g		
Np-237 Pu-239,240 Am-241	10	100	100
Pu-241	350	3500	3500

[1]Total of all radionuclides with less than 5-year half-life.

The 5 $\underline{M}$ Na limit for grout was thought to represent the upper limit of currently demonstrated grout formulations, based on information such as a feed sodium concentration of 122,000 ppm and a volume of grout being 1.4-fold greater than the volume of the feed solution. However, it was subsequently learned that the "ppm" units given for the concentration of sodium in the grout feed were meant to indicate mg/L rather than the generally accepted mg/kg. This means that the currently demonstrated maximum concentration of sodium in grout is really ~4 $\underline{M}$ rather than the 5 $\underline{M}$ value assumed here. However, we will continue to use the 5 $\underline{M}$ assumption to be consistent with Straalsund et al. (1992).

It should be pointed out that the compositional envelopes of suitable grouts have not yet been defined; thus, estimates of the minimum grout volume resulting from processing of tank wastes are subject to large uncertainties. The currently demonstrated grout feed composition "limits" are really not limits at all; they merely reflect the compositions of grout feeds that result from the compositions of wastes studied to date and the grout limitation on radiolytic heat generation.

Also shown in Table 5 are the estimated quantities of some radionuclides that are present in the wastes, but are not governed by the NRC regulations. These are presented for comparison and because one of them, Sm-151, enters into a flowsheet decision. Again, only those radionuclides having half-lives >13 years are included here.

The decontamination factors (DFs) shown in Table 5 were calculated assuming concurrent processing of sludge and supernate, with the volume of LLW being defined by the total sodium content of the two fractions. Other processing approaches would lead to different decontamination requirements unless the determinations can be made on an overall average basis. For example, if sludge and supernate are processed separately and if the sludge contains 10% of the sodium, and if each increment of grout

Table 5. Important radionuclides in SST waste.

Radionuclide	Total Quantity, Ci[1]	Concentration in Average Waste at 5 $\underline{M}$ Na, Ci/m^3	Required Decontamination Factor (DF) for Class A[2]
Class A Listed			
C-14	1.4 x 10^2	3.1 x 10^{-4}	3.9 x 10^{-4}
Ni-63	3 x 10^5	6.2 x 10^{-1}	1.8 x 10^{-1}
Sr-90	4.5 x 10^7	1.0 x 10^2	2.5 x 10^3
Tc-99	9 x 10^3	2.0 x 10^{-2}	6.7 x 10^{-2}
I-129	2.4 x 10^1	5.3 x 10^{-5}	6.6 x 10^{-3}
Cs-137	9.5 x 10^6	2.1 x 10^1	2.1 x 10^1
Np-237	3.1 x 10^1	6.8 x 10^{-5}	4.0 x 10$^{-3(3)}$
Pu-239	2.7 x 10^4	6.0 x 10^{-2}	3.5 x 10$^{0(3)}$
Am-241	3.6 x 10^4	8.0 x 10^{-2}	4.7 x 10$^{0(3)}$
Other			
Se-79	1 x 10$^{1(4)}$	3 x 10^{-5}	
Zr-93	4 x 10$^{3(4)}$	9 x 10^{-3}	
Pd-107	3 x 10$^{1(4)}$	7 x 10^{-5}	
Sn-126	4 x 10$^{2(4)}$	1 x 10^{-3}	
Sm-151	4 x 10$^{5(4)}$	1 x 10^0	
Th-232	1$^{(4)}$	3 x 10^{-6}	
U-238	4 x 10$^{2(4)}$	9 x 10^{-4}	
		1.2 x 10^2	

[1]From values in HDW-EIS (DOE 1987) unless indicated otherwise. Daughter activities not included.

[2]Required DF = (concentration in waste at 5 $\underline{M}$ Na) ÷ (concentration allowed in Class A LLW); see Table 4 for Class A values. A value <1 indicates that no decontamination may be necessary to meet Class A regulations.

[3]For a waste form density of 1.7 g/mL.

[4]Separate estimate.

is required to meet the Class A limits, then tenfold higher DFs would be required while processing the sludge for those radionuclides (e.g., strontium, plutonium, americium) that are present in the sludge.

The data of Table 5 show that only Sr-90, Cs-137, Pu-239, and Am-241 need to be removed from the SST waste (considering an average of all tanks) for the LLW to meet Class A limits. However, removal of other components might be required because of variabilities in compositions among the tanks (e.g., Ni-63). Removal of Ni-63 was not included in the example flowsheet; it will need to be added to ensure that the LLW form does not exceed the Class A limit during processing of some of the SST waste unless the waste feeds can be blended appropriately or unless the radionuclide content of the grout can be considered on an overall average basis instead of on an individual increment basis.

If it is necessary to implement a nickel removal process, then disposition of the nickel becomes a problem because it cannot be accommodated by HWVP-type glass without exceeding the ~1000 canister goal (Table 3). Decay storage in a surface facility or repository disposal in an alternative waste form are possibilities for handling this special problem.

The process steps used to separate the indicated radionuclides from the waste must be capable of providing higher DF values than those given in Table 5 in light of the sum-of-fractions-rule portion of the Class A regulations and feed variability possibilities, especially in the SSTs. Target DFs tenfold greater than those listed in Table 5 may be required to ensure that clean option goals are met.

Because DFs in excess of ~10^4 may be difficult to achieve in practice in one cycle, sequential cycles may be needed in some cases to achieve adequate removals of strontium and cesium. Such needs could be obviated by appropriate blending of feeds or by overall averaging of the contents of the LLW grout, as was discussed earlier for Ni-63. The current flowsheet, based on the overall average feed composition, implicitly assumes that such blending does occur.

ALARA-Driven Separations Requirements. As was discussed earlier, Tc-99, I-129, and U-238 can be prime contributors to long-term radioactivity-related risk even at concentrations that are allowed by the Class A LLW limit. Thus, the example clean option flowsheet includes steps to reduce the concentrations of these radionuclides in the LLW fraction, which will be disposed of under much less stringent conditions than will the HLW fraction.

The separated Tc-99 can be accommodated within 1000 HLW canisters, but not the separated U-238 (Table 3). Thus, it is assumed that the separated uranium will undergo additional purification steps so that it can be added to the existing stockpile of such material. The separated iodine is assumed to be disposed of in a special form, as it could not be easily incorporated in HLW glass.

Other ALARA-driven separations incorporated in the example flowsheet are those to send Th-232 and Sm-151 to the HLW fraction. Thorium separation is included simply because it will likely occur along with the separation of plutonium. Samarium separation is included because Sm-151 is one of the major radionuclides present in the waste (Table 5), and this fission product lanthanide can be separated along with americium from the bulk chemical lanthanides (primarily lanthanum and cesium) that are present in the wastes because of their use in processing steps.

Disposition of Tank Waste Constituents. Table 6 presents a summary of the disposition of the tank waste constituents. As discussed in the previous subsections, most of the bulk chemical constituents of the wastes must go the LLW grout in order to dispose of most of the radioactivity within ~1000 canisters of HLW glass. Grout is considered to be the best demonstrated available technology for toxic metals in non-nuclear industries, and should be applicable to these mixed wastes as well. However, the toxicity of the grout waste form can be reduced by destroying materials such as nitrate, nitrite, and organics. These destructions are assumed in this example flowsheet to be accomplished by calcination; this has the added benefit of allowing the recovery of sodium hydroxide for flowsheet use without affecting the volume of the LLW fraction.

EXAMPLE FLOWSHEET DESCRIPTION

For purposes of this study, an example flowsheet was constructed to provide the separations needed to achieve the clean option objectives, starting with a feed made up of the total SST contents. It is known that processing the waste on a tank-by-tank basis

Table 6. Disposition of tank waste constituents.

LLW	HLW	Destroy	Special Disposal	Stockpile
Hydroxide	Cs	Nitrate	I	U
Carbonate	Sr	Nitrite	Ni[1]	
Phosphate	Tc	Organics		
Sulfate	Th			
Fluoride	Np			
Chloride	Pu			
Ferrocyanide	Am			
Al	HLn[2]			
Ba				
Bi				
Ca				
Cr				
Fe				
Hg				
K				
LLn[3]				
Mn				
Na				
Ni[1]				
Pd				
Si				
Zr				

[1]Nickel is currently planned for LLW disposal, but combinations of processing sequence and regulatory interpretation could necessitate a special disposal form in some cases.

[2]HLn = lanthanides heavier than promethium.

[3]LLn = lanthanides lighter than promethium.

would give a wide range of feed compositions, thus perhaps a wide range of required decontamination needs. In some cases additional cycles may be needed for radionuclides requiring high decontamination factors (e.g., Sr-90) and in other cases it may be possible to eliminate some process steps (e.g., Cs-137 removal from dissolved sludge). However, this flowsheet is thought to provide a reasonable example of the process complexity, and low waste generation capability, of a flowsheet needed to achieve the objectives of the clean option.

The flowsheet includes processes to remove several radionuclides (e.g., Sr-90, Tc-99, Cs-137, and perhaps Pu-239) from both acidic and basic solutions, even though these materials will likely be predominantly in only one type of solution. This was done because of the likelihood that, with some wastes at least, the quantity, though small, of some of these radionuclides present in the "other" solution will still mean that the required overall DF cannot be achieved without treating the "other" solution.

The example clean option flowsheet developed for the preliminary evaluations of technical feasibility and of impact on waste volumes is outlined in Figure 1. This flowsheet was developed with SST wastes specifically in mind, but it should be amenable to DST wastes as well.

A number of potential processes are available for each of the radionuclide removal and separation steps needed to meet the clean option objectives. Similarly, the

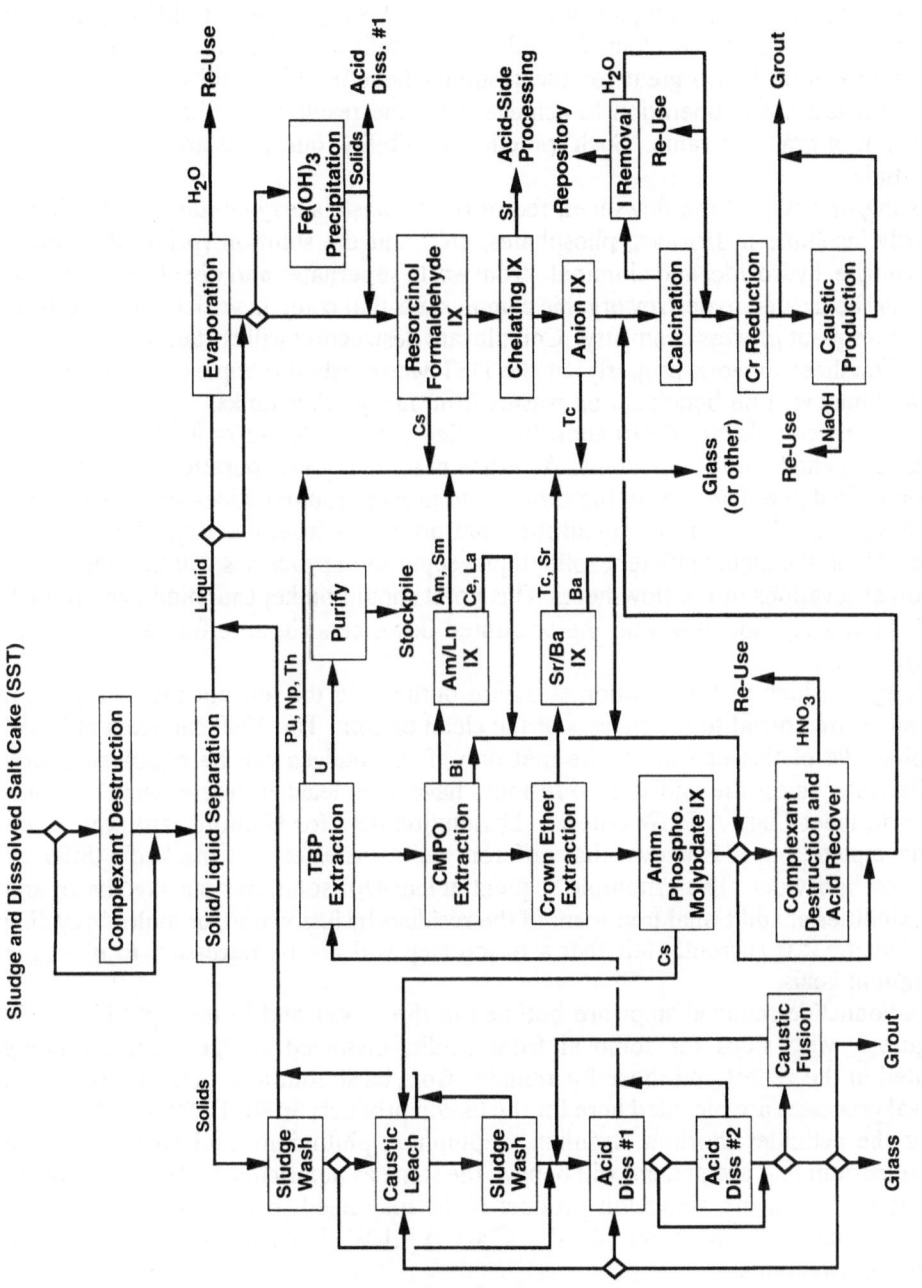

Figure 1. Example clean option overall flowsheet.

individual process steps can be used in a number of different sequences. The processes and the process sequence identified in Figure 1 were selected after consideration of the suitability of options with regard to 1) suitability for the specific problem (Hanford tank wastes), 2) level of development, and 3) ease of developing meaningful estimates of chemical additions to the wastes from published information.

This flowsheet was developed primarily with only "technical feasibility" in mind. The importance of process complexity and costs is well recognized and will be addressed in future efforts. Future cost/benefit evaluations could well conclude that the cost of a certain operation is too great for the resultant benefit; this could lead either to a search for a less costly operation to achieve the same result or to a decision that the result is not of practical value. Such "practical feasibility" questions are not addressed in this study.

As the first step of the flowsheet, the retrieved waste slurry containing the sludge (primarily insoluble hydroxides, phosphates, etc.) and the solution (primarily sodium nitrate/nitrite, hydroxide, and aluminate from waste supernates and dissolved waste salt cakes) may be treated to destroy organic complexants that could have deleterious effects on the subsequent process chemistry. Complexant destruction will certainly be required for this flowsheet to work properly on the DST waste called complexant concentrate (CC), and may well be beneficial on wastes from many other tanks.

Following complexant destruction (if applied), the waste slurry is divided into its solid (sludge) and liquid components. An efficient solid/liquid separation (SLS) process will be required here (or later in the process) to prevent radionuclide-containing solids from entering the LLW form in quantities that prevent it from meeting Class A LLW criteria. Similarly, highly efficient solid/liquid separation processes will be required at many other locations in the flowsheet. This point should be kept in mind even though these locations are not shown on the presented flowsheets (in an effort to keep them simple).

Sludge washing and dissolution steps are outlined in the left portion of Figure 1. These steps are crucial to the success of the clean option. The 1000 canisters of HLW glass objective of this option can be met only if the sludges can be dissolved nearly quantitatively, unless the undissolved residues have been leached free enough of radionuclides to meet Class A LLW criteria. Dissolution data for Hanford tank wastes are very incomplete, but it is known that different tank wastes exhibit markedly different dissolution behaviors. Thus, multiple sequential leach/dissolution steps are shown and the possibilities of additional treatment of the residues by fusion and/or multiple cycling are included. It is currently felt that a fusion step will not be necessary to meet the clean option goals.

Radionuclide removal steps are outlined in the center and in the right-hand side of Figure 1, with steps for removal from acidic, dissolved sludge solutions being presented in the center and those for removal from basic solutions being on the right. Removal processes are included here for the fission products Sr-90, Tc-99, I-129, Cs-137, and for the actinides thorium, uranium, neptunium, plutonium, and americium (the traces of curium that are present will behave the same as americium). As was discussed under Study Bases, the activation product Ni-63 may need to be removed from the wastes in some tanks in order for the Class A LLW limits to be achieved in all processing increments.

In order to meet the clean option objective of incorporating the HLW within ~1000 glass canisters, separations steps beyond the removal of radionuclides from the waste streams will also have to be performed. Two such steps are included in Figure 1: 1) the separation of americium from lanthanide elements that were added as process chemicals, and 2) the separation of strontium from barium.

A major feature of this flowsheet is the recycle of process chemicals (especially sodium hydroxide) in order to minimize the "growth" of waste during processing. Calcination is included in Figure 1 primarily as a means of removing nitrate, nitrite, and organic complexants so that the LLW waste form will be less toxic and will have greater stability. Recovery of sodium hydroxide from the calcined waste oxides should be relatively straightforward.

Sludge Washing and Dissolution

Essentially complete sludge dissolution or, alternatively, extensive leaching of radionuclides from undissolved sludge components, is necessary for success of the clean option. Experimental results have been obtained with several different types of sludges (Lumetta and Swanson 1993; Lumetta et al. 1993); in nearly all cases the clean option objectives were met quite simply. However, studies with a much greater range of tank sludge compositions will be required before a sludge dissolution flowsheet can be defined. The flowsheet given in Figures 2 and 3 provides a good indication of what might be finally determined to be appropriate.

With some wastes, as was mentioned earlier, the first step in the pretreatment series may be a complexant destruction treatment. This is because some wastes are known to contain high enough concentrations of organic complexants that they interfere significantly with the effectiveness of radionuclide removal processes (Schulz 1980). Ozone has long been considered for use in such an application (Schulz 1980), largely because its reduction product does not add any nonvolatile material to the waste. Avoiding the addition of nonvolatile (and insoluble) materials was an important consideration in the earlier approaches to tank waste pretreatment, where sludge vitrification was assumed, but it is of much less importance in the clean option, where sludges will be dissolved and their bulk components will go to LLW grout. Thus, (as suggested by D. O. Campbell, consultant) oxidants such as potassium permanganate might find application here, especially in tanks where the organic complexant concentrations are not high.

One possible side effect of oxidative destruction of organic complexants is that the concentration of plutonium in the liquid phase might be increased by oxidation of Pu(IV) to Pu(VI), which is sorbed less efficiently by the sludge components. If such an effect is too large, the Pu(VI) can be reduced back to Pu(IV) before the liquid is separated from the sludge.

The current volume of SST has been estimated to be 1.4×10^5 m^3 (1.4×10^8 L). It is assumed here that retrieval and/or salt cake dissolution operations will increase this volume fourfold, ending up with a liquid supernate phase having a sodium concentration of 4 $\underline{M}$ and a solid (sludge) phase containing the insoluble metal hydroxides, phosphates, etc.

In the flowsheet portion shown in Figure 2, it is assumed that the SLS step is filtration or centrifugation with the collected solids being washed with sufficient water to increase the liquid volume by 5%. This should give efficient removal of the interstitial liquid components from the solids.

The first dissolution step of the example flowsheet is a caustic (NaOH) leach step aimed at dissolving selected solids (e.g., aluminum hydroxide, silica, silicates) and converting some insoluble phosphates to hydroxides. This step is envisaged as involving digestion at the boiling temperature for several hours in several-molar sodium hydroxide; conditions that have been found to dissolve incinerator ash (Thompson et al. 1979) and fiberglass (Jantzen 1990), materials similar in composition to some of those that may be present in the waste tanks. An initial NaOH concentration of 4 $\underline{M}$ is shown for this step; the indicated volume should be sufficient to prevent the solubility of

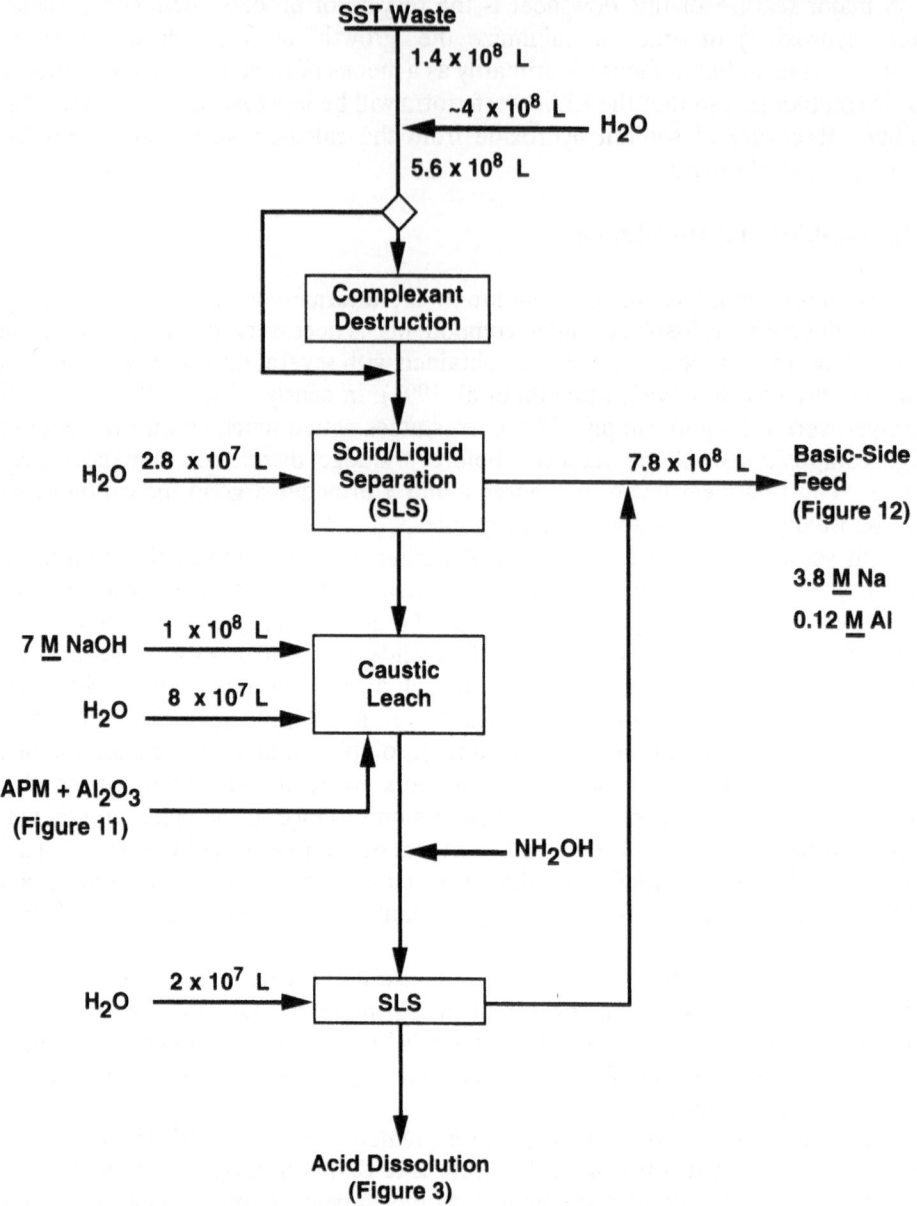

Figure 2. Sludge washing and caustic leaching portion of flowsheet.

sodium aluminate from being exceeded. Data of Delegard (1985) indicate that the room-temperature solubility of Pu(IV)-hydroxide in the resultant leach solution would be <10 nCi/mL; if necessary, this could be decreased by addition of a reducing agent such as hydroxylamine, as is shown in Figure 2.

It is important to note that the NaOH used in this leach step is recovered from a calcine of the decontaminated waste solution (see Sodium Hydroxide Reuse and Disposal), and thus does not add more material to the waste. This is also true of the NaOH used in many other flowsheet steps.

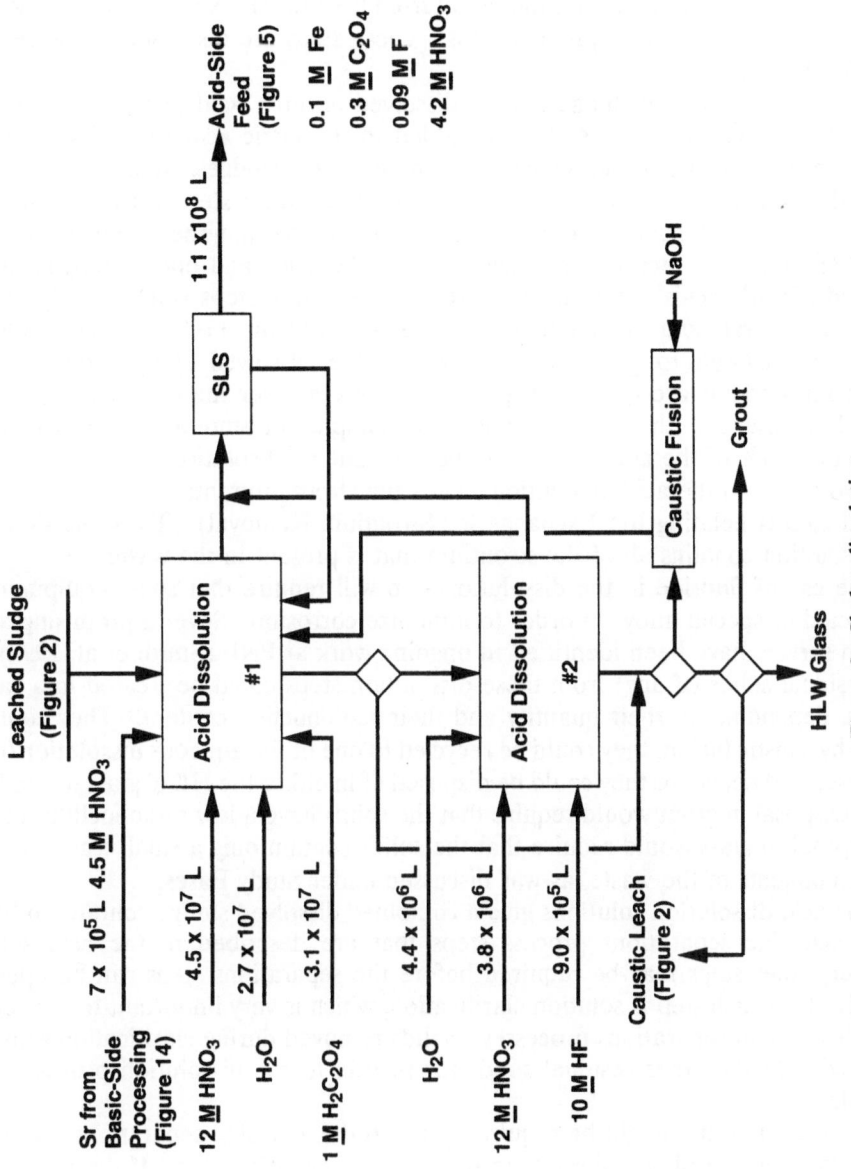

Figure 3. Sludge dissolution.

In addition to the waste sludges, the caustic leach step will also treat the ammonium phosphomolybdate used to remove cesium from acidic solution (see APM Ion Exchange for Cesium Removal). Residues from the oxidation of spent ion exchange resins used in many separations steps will also be treated here.

The solids remaining after the caustic leach step are again collected by filtration or centrifugation, and are washed to remove the interstitial solution. The leach and wash solutions are combined with the liquid from the first SLS step to provide the feed to the basic-side separations process steps. These steps are discussed in the section, Basic-side Separations.

The washed, leached sludge is then dissolved in nitric acid (Figure 3); it is very likely that complexants will need to be added to the nitric acid to achieve efficient dissolution (or leaching of radionuclides) of many of the sludges. It is also very likely that, with some wastes at least, more than one dissolution step will be required to achieve complete dissolution (or leaching). Thus, the flowsheet shows two acid dissolution steps, one employing mixed nitric-oxalic acids and one employing mixed nitric-hydrofluoric acids. With many waste sludges, these steps will probably involve periods of several hours and temperatures at or near the boiling point. Solution volumes are chosen to give ~0.1 $\underline{M}$ iron in the dissolved sludge solution, and complexant amounts are chosen to give a slight excess over the quantity required to form 1:1 complexes with the metal ions. The complexant amounts given in Figure 3 assume that 10% of the aluminum is in the acid-side feed solution.

Also added to the acid dissolution step is the strontium removed in the basic-side removal step (Chelating Ion Exchange for Strontium Removal). Thus, the dissolved sludge solution contains all of the strontium that is present in the waste.

The use of fluoride in the dissolution step will require that some equipment be constructed of special alloys in order to minimize corrosion. Several promising alloys for such service have been identified in ongoing work at PNL (Smith et al. 1992).

Residual solids (if any) from these dissolution steps could be treated in a variety of ways, depending on their quantity and their radionuclide content. They could be treated by caustic fusion; they could be recycled to one of the aqueous dissolution/leach steps discussed above; or they could be disposed of in either the HLW glass or the LLW grout. Disposal in grout would require that the solids have a low radionuclide content, and disposal in glass would require that the solids contain only a small fraction of the bulk components of the waste, as was discussed under Study Bases.

The acid dissolution solutions give a combined dissolved sludge solution to be fed to the acid-side separations process steps that are described in the next section. However, other steps may be required before the separations steps can be operated properly. One such step is solution clarification, which is very important to the success of highly efficient separations processes. Solids removed during clarification would be combined with the other residual solids for recycle to the dissolution process or for disposal.

Another step that might be required prior to the acid-side separations processing comes from the use of complexants in the sludge dissolution step. If the amounts of complexants are too much in excess of the complex-forming bulk metal ions present, they may complex significant portions of the actinides and thus decrease their efficiency of removal.

If necessary, oxalate can be destroyed by digestion of the dissolved sludge solution, especially in the presence of manganese ion (Bibler et al. 1981). Manganese is present in SST sludge (Table 1), and more could be added to give a more rapid reaction without increasing the waste volume because manganese will be only a minor component of the LLW grout. In fact, the use of permanganate to rapidly destroy the oxalate could be appropriate (but is not included in the current flowsheet).

A class of substituted diphosphonic acids that are stronger complexants than oxalate but are more easily destroyed by reaction with nitric acid have been termed TUCS (for thermally unstable complexants) compounds by workers at Argonne National Laboratory (Horwitz et al. 1990). These compounds have some advantages over oxalate, but they suffer from the potential disadvantage that phosphoric acid is a product of their destruction. Phosphoric acid should not present any problem in LLW disposal, but there is concern that insoluble phosphates might form and cause problems in some steps of the flowsheet. The effects of these compounds should be considered in more depth before their use is adopted.

Fluoride cannot be destroyed like organic complexants, but its complexing of actinide ions can be minimized by providing additional amounts of complex-forming metal ions (e.g., Al^{3+}, Zr^{4+}) to tie up most of the fluoride. With the added fluoride concentration and the overall average SST zirconium concentration shown in Figure 3, most of the fluoride would be tied up even if no additional metal were added. However, fluoride might be used in the dissolution of some sludges that do not contain either aluminum or zirconium; in those uses, addition of some complex-forming metal ion would be necessary for efficient actinide extraction. If aluminum is used for this purpose, as it was for years when fluoride-containing feeds were processed in the Hanford PUREX plant, the quantity of aluminum needed (for the amount of fluoride given in Figure 3) would amount to only ~10% of that already present in the waste. Thus, such a use should have no effect on the quantity of grout required for the LLW fraction of the waste.

Acid-Side Separations

The (acidic) dissolved sludge solution is expected to contain nearly all of the actinides and the strontium, as well as nickel and other transition metals (e.g., iron, chromium) that are present in the wastes. It may also contain small, but significant, fractions of the cesium, technetium, and aluminum. An acidity of ~4 $\underline{M}$ is assumed for the feed to the acid-side separations processes, primarily because of the feeling that attainment of "complete" sludge dissolution will be easier at an acidity that high, and of the desire to avoid dilution of dissolved sludge solutions.

Radionuclides that need to be removed from the dissolved SST sludge solution in order to meet Class A LLW criteria are Sr-90, Cs-137, Pu-239, and Am-241; Ni-63 removal may also be required when processing some SSTs, as was discussed under Study Bases. Other radioactive isotopes whose removal would be potentially beneficial from the standpoint of long-term risk are Tc-99, I-129, Np-237, and U-238 (Buck and Peffers 1991; Droppo et al. 1991). Demonstrated or highly promising processes exist for removing all of these radionuclides (with the possible exception of Ni-63 and I-129) from dissolved sludge solution. Process selection and process sequencing require consideration of many factors; some of those considerations will be discussed next. It should be emphasized that most of these factors were considered in only a qualitative fashion, and many decisions were based on then current knowledge and the availability of data required to describe an example flowsheet.

The processing sequence selected here is to first remove the actinide elements from the dissolved sludge solution and then remove the fission products. This sequence was chosen to ensure that the transuranic actinides would be removed before they could contaminate the solvent and/or sorbent material used in fission product removal processes, and perhaps thus present processing complications.

Figure 4 presents distribution coefficient (D) values for several important actinide (and other) constituents of Hanford SST wastes between nitric acid solutions and a) 1.1 $\underline{M}$ (30%) TBP-NPH, the solvent used for years in the PUREX process, and

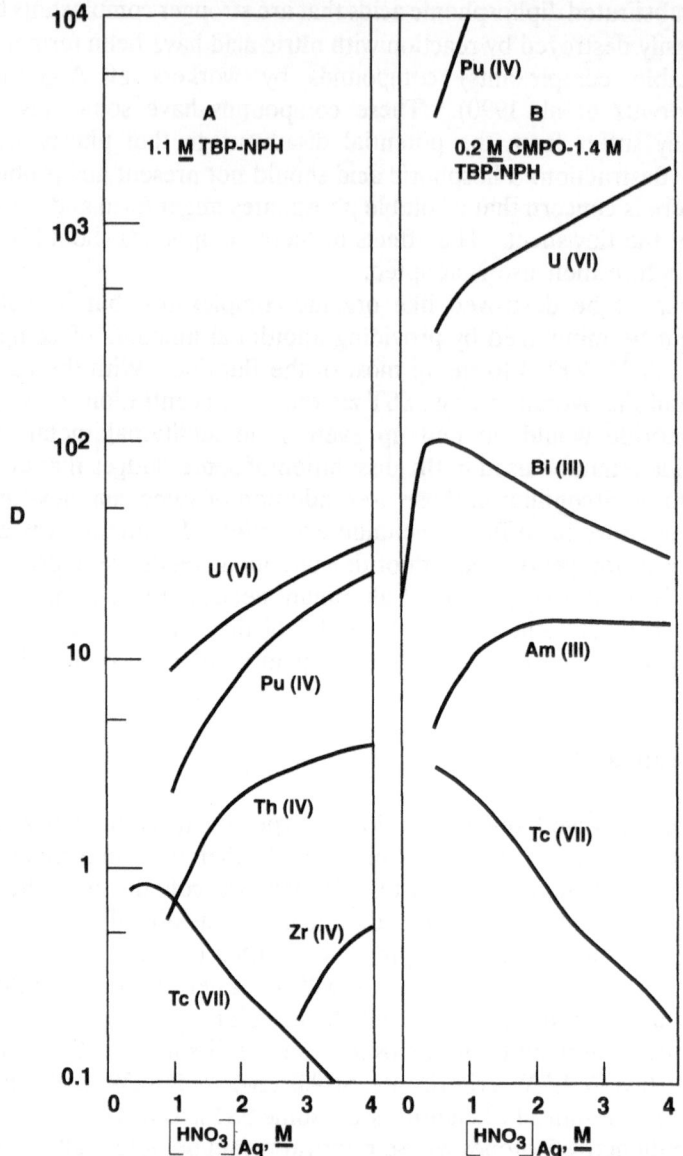

Figure 4. Typical distribution coefficients for TBP and CMPO extractions.

b) 0.2 $\underline{M}$ CMPO - 1.4 $\underline{M}$ TBP-NPH, a typical solvent composition proposed for the TRUEX process. (D is defined as the concentration of a material in the organic phase divided by its concentration in the aqueous phase.) The bismuth data shown here were obtained from Lumetta et al. (1993); the other data were obtained from Bond (1990) and from Horwitz et al. (1985). These data are shown to compare the relative strengths of the two extractants, and to provide some background for the following discussion. They were obtained under nonprocess-type conditions (e.g., absence of macro amounts of uranium, which tends to load the solvent and thus decrease the D values, and absence of complexants, which also decrease the D values), and should not be used in a quantitative way.

Earlier studies of the removal of actinide elements from dissolved Hanford sludge solutions have centered on the use of the TRUEX solvent extraction process to simultaneously remove all of these elements (and sometimes technetium). The actinides (and coextracted lanthanides) would be disposed of in the HLW glass, and the bulk sludge components, which do not extract, would be disposed of (after a strontium removal process) in the LLW grout. In a more recent approach, the strontium and actinide removal processes would be combined in one step (Horwitz 1991). The degree to which the earlier studies considered the behavior of bismuth is uncertain.

Bismuth is present in SST sludges in significant quantities (Table 1) because of the use of a bismuth phosphate carrier precipitation process in the first plutonium recovery plants. This quantity may be too large to be accommodated in 1000 canisters of HLW glass (Tables 2 and 3), so separation of bismuth from the TRU elements is planned. However, bismuth exhibits an acid dependency in its extraction by TRUEX solvent (Figure 4b) that may complicate its separation from the TRU elements.

Several options have been explored in the earlier studies for stripping extracted actinides from the TRUEX solvent. Trivalent actinides can be stripped into dilute acid, but the high distribution coefficients of the tetravalent and hexavalent ions make it necessary to use complexants to strip them. Two principal stripping approaches have been proposed: 1) simultaneous stripping of all actinides using complexing agents, and 2) selective stripping of a) trivalent actinides (e.g., americium) into dilute acid; b) tetravalent actinides (e.g., plutonium) into dilute acid containing fluoride as a complexing agent; and c) stripping hexavalent actinides (e.g., uranium) into carbonate solution. In these approaches, extracted technetium is stripped only by carbonate solution.

If Approach 1 were used for the clean option with SST waste, additional processing of the stripped mixture would be required to separate uranium (and possibly bismuth) from the transuranic elements (if such processing were not done, the amount of glass required to contain the TRU-contaminated fraction would exceed 1000 canisters). Additional processing might also be required to separate the TRUs from the complexing agents; this would be the case if the complexing agents (or their decomposition products) interfere in the vitrification process. Approach 2 would require additional processing to recover uranium (and perhaps technetium) from the carbonate solvent wash solution, and the uncertain behavior of bismuth could be a complicating factor.

Considerations such as these led to the decision that, while the above approaches are certainly feasible, it would be simpler to prepare a defensible example flowsheet employing first a TBP solvent extraction cycle to handle the recovery and needed purifications of the tetravalent and hexavalent actinides and then a CMPO extraction cycle to handle the recovery of the trivalent actinides and their separation from bismuth. This approach also eliminates the use of complexants in stripping actinides. Such a flowsheet may also be simpler to operate, but that will not be known until much more detailed study has been completed. In addition, high acid TBP and CMPO cycles were selected for the example flowsheet to minimize technetium extraction and ensure that the carbonate wash solutions used for solvent purification would not contain Tc-99 in amounts that exceed the Class A limit.

Extractants other than CMPO may also be suitable for use in extracting the trivalent actinides. CMP (also know as DHDECMP) was considered for use in this example flowsheet; one reason for not selecting it was the absence of bismuth extraction data [other than the "partial retention" noted by Marsh and Yarbro (1988) in an extraction chromatographic study] on which to base an example flowsheet. The pros and cons of these two extractants should be evaluated in more depth before any choice for final implementation is made, but the ultimate choice should have little, if any, effect on conclusions based on the example flowsheet using CMPO.

Thus, the processing sequence chosen here for the acidic, dissolved sludge solution is: 1) a TBP solvent extraction step to remove tetravalent and hexavalent actinides (pentavalent neptunium is converted to the hexavalent state for removal here also), 2) a CMPO solvent extraction process for removal of trivalent actinides (trivalent lanthanides are also removed), 3) a crown ether solvent extraction process for removal of strontium and technetium, and 4) an ammonium phosphomolybdate (APM) ion exchange process for removing cesium. These steps are discussed further in the following sections.

TBP Solvent Extraction for Removal of Uranium, Plutonium, Neptunium, and Thorium. The TBP solvent extraction process for removal of uranium, plutonium, neptunium, and thorium is shown in Figure 5. For a perspective regarding plant size, it will be mentioned that the feed flow rate for processing the dissolved SST sludges in 20 years of operation would be approximately the same as the feed flow rate in a spent fuel reprocessing plant that processes 800 MTU/year (which is the capacity of the plants recently completed in France and planned for construction in Japan).

The feed to this cycle contains not only the dissolved sludge solution (Figure 3), but also a stream to provide chemicals to adjust (if necessary) the degree of complexation of the contained ions, and a stream containing the bottoms solution from the backcycle evaporator used to concentrate the waste solution from the subsequent uranium purification cycle. The latter two streams are discussed later in this section; for now, the proper degree of complexation is assumed to be achieved in the sludge dissolution step so that no chemicals (other than water) are used in the complexation adjustment solution.

In the first extraction contractor of this TBP cycle, uranium, plutonium, neptunium, and thorium are extracted into the TBP organic phase, which is then scrubbed with nitric acid to increase the separation from the bulk metal ions, some of which have a slight extractability. The raffinate from this contractor contains the remaining components of the dissolved sludge solution, and is routed to the next step of the treatment process (following section). The extracted plutonium, neptunium, and thorium are selectively stripped back into an aqueous phase in the second contractor (to be routed to HLW glass for disposal); and the uranium is stripped in the third contractor. The uranium is purified further in another TBP cycle so that it can be added to the existing uranium stockpile.

The solvent composition chosen here is the same, 30% (~1.1 $\underline{M}$) TBP in hydrocarbon diluent, as is usually used in the well-known PUREX process. At the ~4 $\underline{M}$ HNO_3 concentration of the feed and scrub streams, efficient removal of uranium, plutonium, and thorium can be achieved. Both the high acidity and the relatively low saturation of the TBP (0.2 $\underline{M}$ uranium) assist in efficient recovery of neptunium by the nitrite-catalyzed oxidation of inextractable Np(V) to extractable Np(VI), as was done for years in the Hanford PUREX plant. The dissolved sludge stream is assumed to be decontaminated from uranium and plutonium by factors of 10^4 and from thorium and neptunium by factors of 10^2. The lower factor is assumed for neptunium and thorium because neptunium removal is controlled by kinetic factors and because thorium is less extractable than uranium or plutonium. Cesium, strontium, americium, and trivalent lanthanides are virtually inextractable by TBP, and technetium extracts very little at the high acidity used here (Figure 4). Thus, these elements will remain in the dissolved sludge stream in this step, along with the nonradioactive elements present in the stream.

Neptunium removal is not needed in order for the LLW to meet Class A limits (Table 5), but is included in the flowsheet for ALARA reasons (and because neptunium tends to be the most environmentally mobile TRU element). Neptunium recovery may

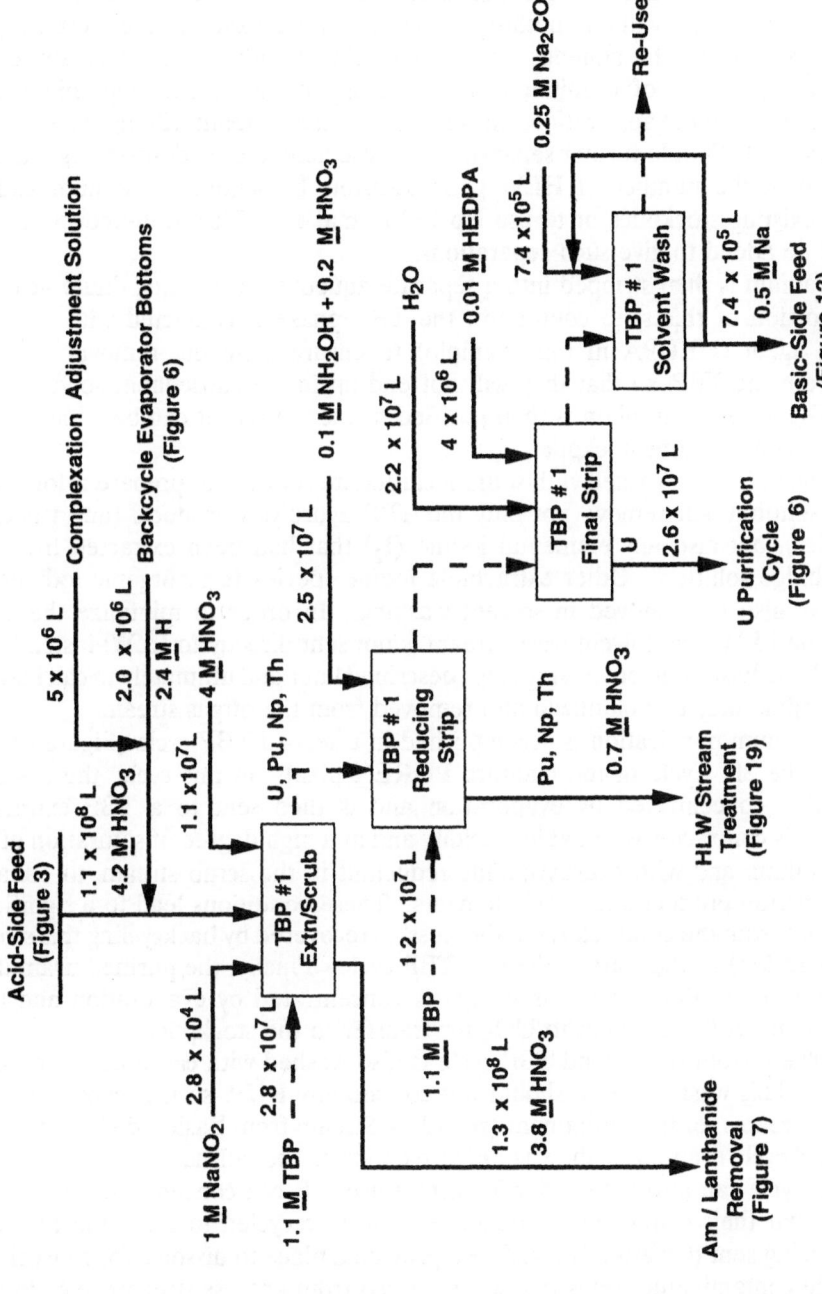

Figure 5. TBP solvent extraction for removal of uranium, plutonium, neptunium, and thorium from dissolved sludge solution.

be beneficial for other reasons (e.g., Pu-238 heat source production), but that is not the reason for removing it from the waste stream in this flowsheet.

The extracted plutonium, neptunium, and thorium are stripped from the TBP phase into an aqueous phase under conditions (~ 0.7 $\underline{M}$ HNO_3 containing reductant) that leave the uranium with the TBP. A reductant is added to reduce Pu(IV) to Pu(III) and Np(VI) to Np(V) to allow these elements to be efficiently stripped. Hydroxylamine (NH_2OH) should be suitable for this purpose; and because it can be decomposed to gaseous products, its use will add nothing to the final waste form. In the "bottom" part of the strip contractor, the aqueous stream is scrubbed with a fresh TBP stream to improve the separation of uranium from the stripped plutonium, neptunium, and thorium. In this example application, the stripped actinide stream will be routed to the HLW for disposal. Should further separation of these actinides be desired (e.g., further minimization of the number of HLW glass canisters by sending plutonium and/or thorium to existing stockpiles or to use Np-237 to make Pu-238), well-known process steps could be added to give such separations.

The uranium is then stripped into a separate aqueous phase using dilute acid. In the "top" portion of this strip contractor, the TBP phase is contacted with a strong complexing agent (HEDPA in this example) to ensure complete removal of TRU elements from the TBP, so that they will not end up in the subsequent solvent wash solution. The stripped uranium is then purified further so that it can be added to the existing depleted uranium stockpile.

The solvent is then washed with sodium carbonate solution to prepare it for reuse. This wash solution will remove not only the TBP hydrolysis products (and traces of radionuclides), but also any elemental iodine (I_2) that had been extracted from the dissolved sludge solution. Other extractable iodine species (e.g., organic iodides), if present, may also be removed in solvent washing. In order to minimize the I-129 content of the LLW, this solvent wash stream is not sent directly to LLW; instead, it is sent through the basic-side processing steps described later and ultimately to calcination, where the iodine may be volatilized and removed from the offgas stream.

Final uranium purification is accomplished in a second TBP cycle (Figure 6) that is based on the 2D Cycle of the Hanford PUREX plant. In this cycle, the uranium stream is first concentrated by evaporation and is then sent to a TBP extraction contractor; this contractor is run at low acidity and at a high degree of saturation of the TBP by uranium, and with hydroxylamine reductant in the scrub stream, to enhance removal of fission product and TRU elements. These conditions lead to a significant uranium "loss" from the contractor, but the "loss" is recovered by backcycling this stream (after evaporation) to the start of the first TBP cycle. Finally, the purified uranium is again stripped into dilute acid; this stream is concentrated by evaporation and then calcined to convert the uranium to UO_3 for transfer to the stockpile.

The solvent from the second TBP cycle is also washed with carbonate to prepare it for reuse. This wash solution should not contain any I-129, so it is combined with other LLW streams for final treatment (see LLW Stream from Basic-side Separations).

The backcycle evaporator shown in Figure 6 to treat the raffinate from the uranium purification cycle can also have other important uses. It can concentrate off-standard streams so that their contained radionuclides will be recycled to the removal cycles, instead of being sent to waste. It could also provide a place to dissolve for re-work any radionuclide-contaminated solids that are removed from process streams (e.g., in feed clarification steps preceding the various separations processes).

For successful operation of the primary TBP extraction step, where most actinides (but nothing else) are removed from the dissolved sludge solution, it is assumed to be necessary to have complexants present in the feed and/or scrub solutions. One reason

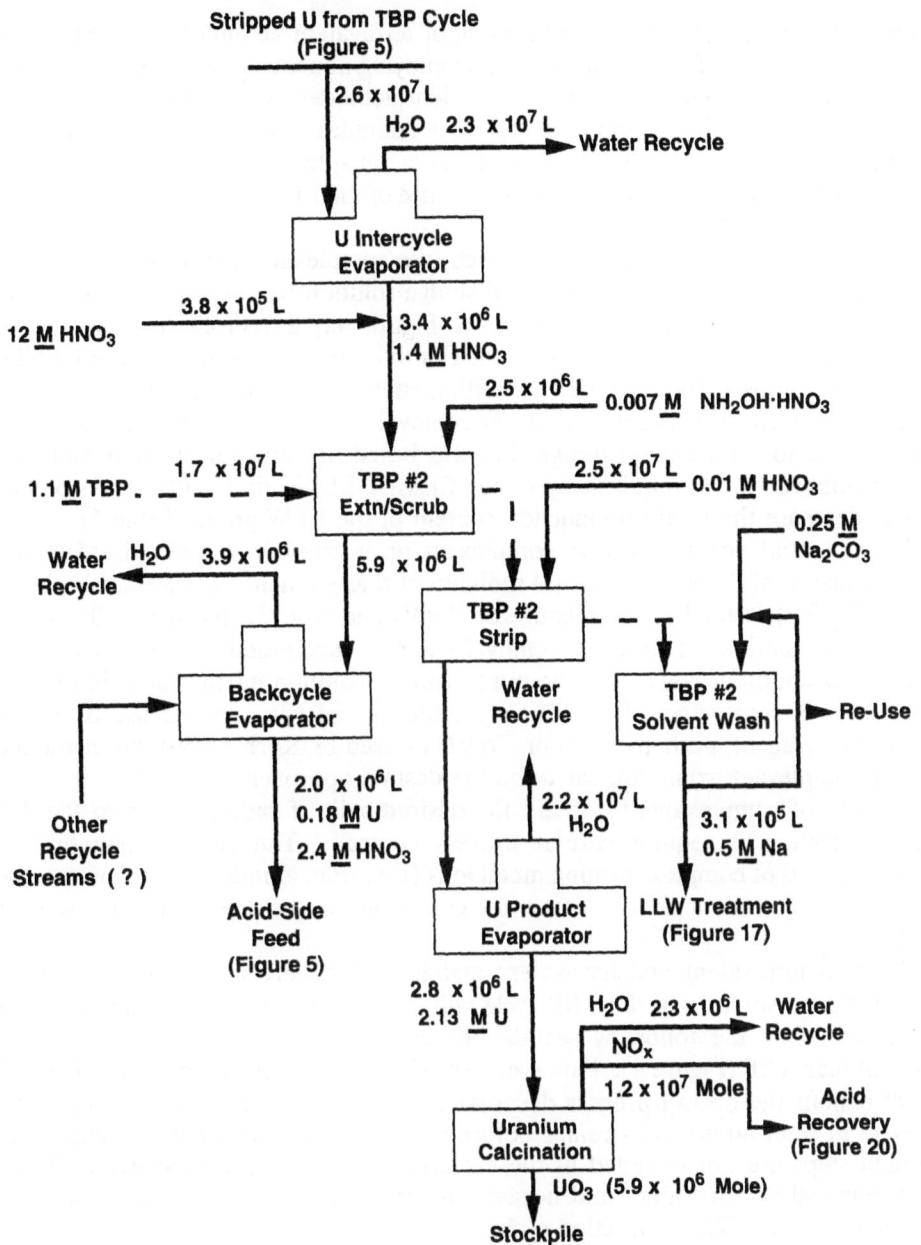

Figure 6. Second TBP extraction cycle for purification of uranium.

is that because of the (relatively) high concentration of zirconium in dissolved SST sludge solutions, highly extractable, mixed Tc-Zr species could form unless the zirconium is "tied up" in a complex.

This effect of zirconium on technetium extraction was investigated by Vialard and Germain (1986), who found D_{Tc} to increase very rapidly as the organic-phase zirconium concentration increases. The D_{Tc} was found to be ~1 from 4 $\underline{M}$ HNO_3 when the organic-phase zirconium concentration was 0.005 $\underline{M}$ (Figure 4a indicates a value <0.1 in the absence of zirconium). However, by adding oxalate to complex zirconium and thus suppress its extraction, these authors were able to decrease the D_{Tc} greater than tenfold without decreasing the extraction of Pu(IV). Because complexing of hexavalent actinides is less pronounced than complexing of tetravalent actinides, it appears to be safe to assume that a TBP extraction cycle employing aqueous-phase complexants can be used to extract the hexavalent and tetravalent actinides away from the technetium and zirconium (as well as from the trivalent actinides and lanthanides and other components of the dissolved sludge solution). Aqueous-phase complexants would likely be present in the feed to this cycle anyway, because of their use in the sludge dissolution process (Figure 3).

If the TBP cycle feed contains too much free complexant, the degree of removal of tetravalent actinides from the dissolved sludge solution will be lowered. Based on the distribution coefficient values given in Figure 4a, a tenfold lowering of the distribution coefficients would still allow Pu(IV) to be removed efficiently, but Th(IV) would not be removed (because its distribution coefficient would be <1). This would not be of great concern because thorium removal was included on the basis of its expected behavior in a system design that was based on removal of other elements. Thorium removal is not required by either Class A LLW criteria (Table 4) or by a desire to decrease the total radionuclide content of the LLW grout (Table 5).

The potential effect of excess complexant on neptunium removal is of greater concern because of the environmental mobility of that element. If excess complexant stabilizes Np(IV) instead of the mixture of Np(V) and Np(VI) that is usually present in nitric acid solutions, neptunium removal could be decreased because of the slow kinetics of oxidation of Np(IV) by the nitrate/nitrite solution mixture used in the TBP cycle extraction contractor. This difficulty could be overcome by the use of a more rapid oxidizing agent, perhaps V(V) or Cr(VI) as used by Koch (1969), but avoidance of excess complexant would appear to be the desired approach.

The use of complexants to obtain the desired split of radionuclides in the TBP extraction cycle will require careful process control. The relative quantities of complexants and of complex-forming metal ions (e.g., iron, aluminum, zirconium) must be properly adjusted, and the extraction conditions to be used must be selected carefully.

Residual tetravalent and hexavalent actinides that were not removed from the dissolved sludge solution in the TBP cycle will be removed in the subsequent CMPO cycle described in the following section because CMPO is a much more powerful extractant than TBP (Figure 4). However, removal of these species in the CMPO cycle will not benefit the overall process decontamination factor unless steps are added to that cycle to treat additional streams before they are routed to LLW. Because such treatment steps are not included in the example flowsheet developed here, efficient overall removal of tetravalent and hexavalent actinides in this flowsheet depends on their removal in the TBP extraction cycle.

CMPO Solvent Extraction and Americium/Lanthanide Ion Exchange. The raffinate from the first TBP cycle provides the feed to the CMPO extraction contractor (Figure 7). The solvent composition assumed for this step is 0.2 $\underline{M}$ CMPO + 1.4 $\underline{M}$ TBP in hydrocarbon diluent, which has been studied extensively by many workers (e.g., Horwitz et al. 1985). This CMPO step is used to remove trivalent actinides (e.g., americium and curium) from the dissolved sludge solution. Trivalent lanthanides are also highly extracted and must be separated in subsequent operations to allow americium to be disposed of in the HLW glass without exceeding the 1000-canister objective of the clean option (Table 3). Depending on the degree of complexation of

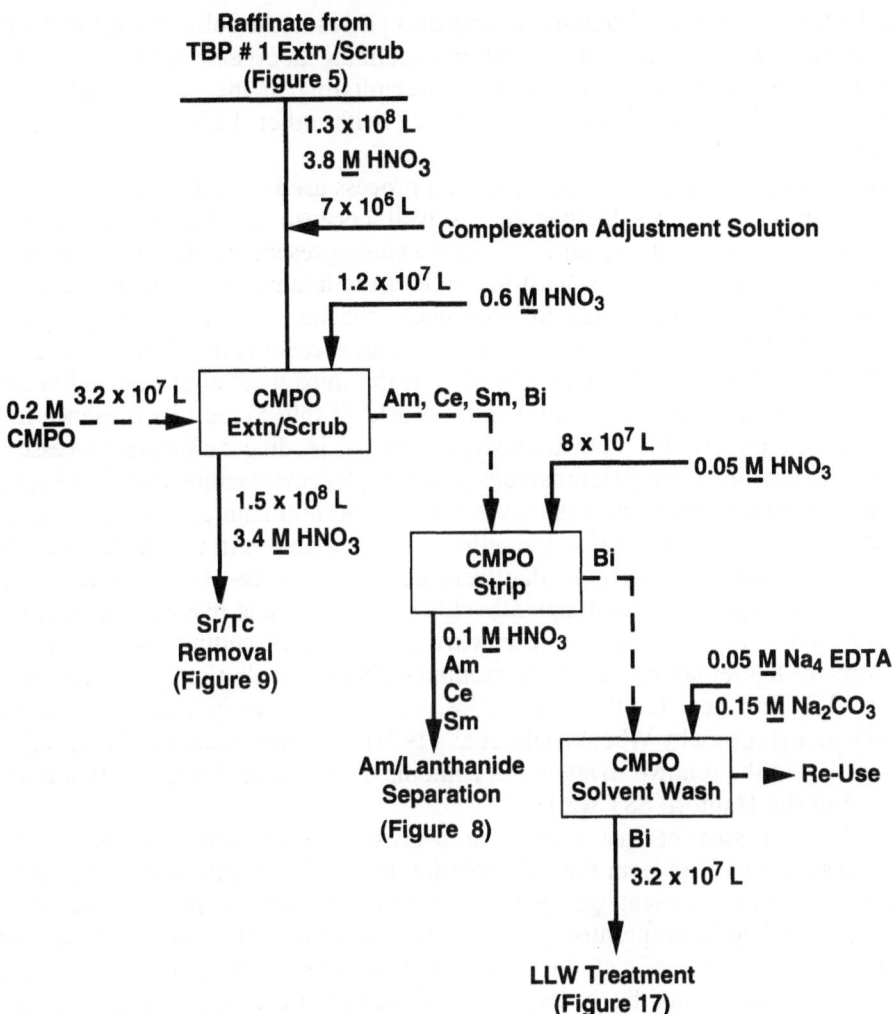

Figure 7. CMPO solvent extraction cycle for removal of americium (and lanthanides) from first TBP cycle raffinate.

the feed, an excessive amount of bismuth may also be extracted; thus, the conditions of the strip steps were chosen to give separation of bismuth from the americium and the lanthanides.

As in the TBP cycle, technetium extraction is minimal at the relatively high feed acidity used here, and strontium and cesium, and the nonradioactive elements in the sludge, are inextractable. The dissolved sludge stream is assumed to be decontaminated from trivalent actinides and lanthanides by factors of 10^4 in this step; however, the lighter lanthanides are subsequently added back to the nonradioactive components (after they have been separated from americium and the heavier lanthanides).

The trivalent americium and lanthanides are stripped from the CMPO extract into dilute nitric acid in the second contractor. This stripping must be at least 99.99% complete in order to maintain an overall americium decontamination factor of 10^4; many strip stages may be needed. The portion of the bismuth that extracted is not well stripped under the indicated conditions, and thus passes to the solvent wash step where it is removed by a sodium carbonate solution containing an organic complexant (EDTA in this example) to prevent bismuth from precipitating in the basic solution. The bismuth-containing wash solution is combined with other LLW streams for final treatment.

As indicated above, the CMPO extraction process used to remove americium from the waste stream removes lanthanides as well because of their similar chemical properties. Because of the quantity of lanthanides present in the waste, additional separation of americium from at least the bulk of the lanthanides must be made if the americium is to be incorporated in 1000 glass canisters (Table 3). The separated lanthanides could all be disposed of in grout without exceeding the Class A LLW limit; however, the lanthanide radionuclide Sm-151 is the third most abundant radionuclide (of those having half-lives >20 years) in the waste (Table 5), and its presence in the grout would be undesirable. Fortuitously, one of the leading candidate processes for separating americium from all lanthanides (band displacement cation exchange) can just as easily be used for separating americium plus heavier lanthanides (such as samarium) from lighter lanthanides. Because the bulk of the lanthanides in the wastes are lighter ones (e.g., lanthanum and cerium) that were added as process chemicals, this separations process allows the americium/lanthanide split to be made in a way that allows the americium and heavier lanthanides to be vitrified without exceeding 1000 canisters and also allows the lighter lanthanides to be grouted without contributing to the radionuclide content. The flowsheet for this separations process is shown in Figure 8; it is based primarily on that given by Wheelwright et al. (1974), with the quantities being adjusted in proportion to the relative trivalent ion contents of the waste studied by Wheelwright et al. and of the Hanford SST waste.

In the first step of this separation process, the trivalent actinides and the lanthanides are removed from the strip solution of the first CMPO cycle (Figure 7) by sorption onto a cation exchange resin. The effluent from the sorption step can be routed to a backcycle evaporator (possibly the one shown in Figure 6) if needed to ensure the maintenance of high decontamination factors to the LLW form, but this is currently assumed to not be necessary. Following a water wash of the resin, the trivalent ions are eluted with a pH-adjusted 0.05 $\underline{M}$ DTPA solution onto a zinc (or other barrier ion)-loaded cation exchange resin column. Continued elution through a series of columns results in the establishment of discrete bands of metal ions in a sequence that depends on the magnitudes of the constants governing the formation of the metal ion-DTPA complexes. This order is zinc; curium, americium, terbium, and dysprosium (together); gadolinium; europium; samarium; yttrium; and then the light lanthanides. For the application in this flowsheet the first effluent cut would contain most of the zinc; the second cut would contain the remainder of the zinc and at least some of the yttrium along with all of the americium, curium, and lanthanides heavier than promethium; and the third cut would contain the remainder of the yttrium plus the lanthanides lighter than samarium.

Final removal of light lanthanides from the resin is accelerated by using a more highly concentrated DTPA solution to strip them. The resin beds are then regenerated, following a water wash, to prepare them for the next cycle. The first (sorption) bed is regenerated with HNO_3, and the subsequent (band displacement) beds are regenerated with $Zn(NO_3)_2$. After another water wash, the resin beds are ready for the next cycle of operation.

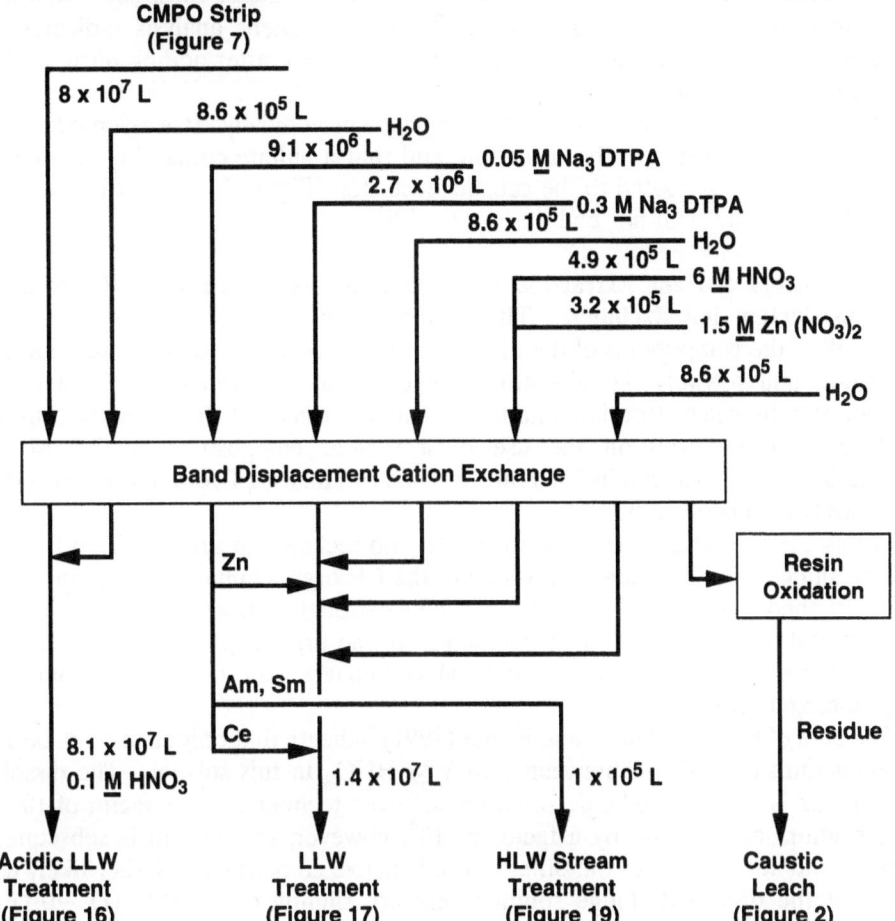

Figure 8. Band displacement cation exchange for separation of americium and heavy lanthanides from light lanthanides.

For this example, pH adjustment of the DTPA solution is assumed to occur by addition of NaOH, and 3 NaOH per DTPA are assumed required. Sodium hydroxide was chosen for use here instead of the ammonium hydroxide proposed by Wheelwright et al. (1974) in order to minimize the potential for release of ammonia gas, which is an environmental concern.

The first and third effluent cuts and the resin stripping and regeneration cuts will be sufficiently free of radionuclides that they can be combined with other LLW effluents, as discussed in Treatment of LLW Streams Resulting from Separations. The second cut could be processed further to separate the americium and heavy lanthanides from the added chemicals (zinc, sodium, and DTPA). However, preliminary analysis indicates that the added chemicals will be present in quantities small enough that they can be accommodated within ~1000 canisters of HLW glass; thus, such additional processing is not included in the flowsheet. If subsequent analysis indicates that additional processing is needed, it could be accomplished, after adding nitric acid, by a second CMPO extraction and strip cycle.

The ion exchange resin will have to be replaced periodically. It is assumed that the spent resin will be destroyed by oxidation, and that a sulfate-containing residue will result. This residue is routed to the caustic leach step (Figure 2), so that its contained radionuclides (if any) will not end up in the LLW.

Crown Ether Solvent Extraction for Strontium and Technetium Removal and Strontium/Barium Ion Exchange. The raffinate from the first CMPO cycle, which contains all of the components of the dissolved sludge solution except the actinides, the lanthanides, and possibly the bismuth, is then fed to a crown ether (CE) solvent extraction step in which strontium and technetium are removed. The flowsheet for this step (Figure 9) is based on the use of a solvent composition of 0.2 $\underline{M}$ di-t-butylcyclohexano-18-crown-6 in 1-octanol, as used by Horwitz, Dietz, and Fisher (1991) in their SREX process study.

In this CE extraction cycle, the strontium and technetium are extracted from the dissolved sludge solution aqueous phase into the CE organic phase; the organic phase is then scrubbed with ~1 $\underline{M}$ HNO_3 to remove less strongly extracted materials; and the strontium and technetium are then stripped from the organic phase into a dilute acid aqueous phase. Also stripped are barium, which behaves very similarly to strontium in this system, and nitric acid.

The data of Horwitz, Dietz, and Fisher (1991) indicate that only strontium, barium, and technetium have D >1 between 1 to 3 $\underline{M}$ HNO_3 in this solvent. The dissolved sludge stream is assumed to be decontaminated from technetium by a factor of 10^2 and from strontium and barium by a factor of 10^4; however, the barium is subsequently separated and added back to the other nonradioactive components. Other likely components of the dissolved sludge solution that are slightly extractable (D ~0.1) are sodium, calcium, molybdenum, ruthenium, and palladium; the portions of these metals that do extract should be easily removed in the scrub section of the contractor. Iron and aluminum are essentially inextractable (D $<10^{-3}$) in this system.

The data of Table 3 indicate that barium must be separated from strontium, if strontium disposal within 1000 canisters of HLW glass is to be achieved. As was discussed previously, the barium content of SST waste given in Table 1 may be too high by a factor of ~10. If this factor were much higher, no strontium/barium separation step would be required. However, such a step is included here (Figure 10) for conservatism.

The method chosen to illustrate the strontium/barium separation is a band displacement cation exchange process that is very similar to that described in the preceding section for separation of the americium and heavy lanthanides from the light lanthanides. In this application, the strontium and barium are removed from the strip solution of the CE cycle by sorption onto a cation exchange resin. The effluent from this sorption step will contain the technetium (which is present in an anionic form); this stream is concentrated and the technetium is routed to HLW. The strontium and barium are then eluted by a complexant solution onto a barrier ion-loaded cation exchange resin column; Bray et al. (1964) reported results with the complexants EDTA and HEDTA and with several different barrier ions. The materials used in this example (Figure 10) are EDTA complexant and zinc barrier ion. The EDTA is actually partially neutralized to give the required pH; it is assumed here that this neutralization is done with sodium hydroxide and that 3 NaOH per EDTA are required. Sodium hydroxide

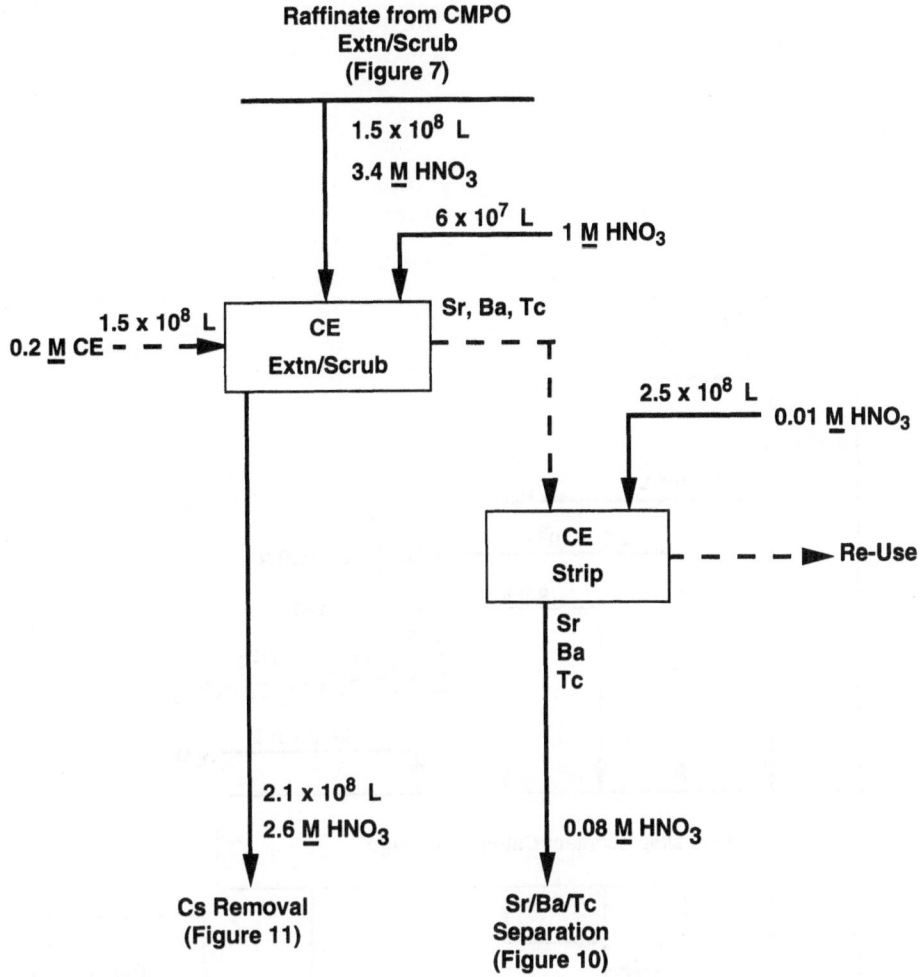

Figure 9. Crown ether solvent extraction cycle for removal of technetium and strontium (and barium) from CMPO cycle raffinate.

was chosen for use here instead of the ammonium hydroxide used by Bray et al. (1964) in order to minimize the potential for release of ammonia gas, which is an environmental concern.

Continued elution of the strontium and barium through a series of columns results in segregation into discrete bands. For the application in this flowsheet, the first effluent cut would contain most of the zinc; the second cut would contain the remainder of the zinc plus all of the strontium plus some of the barium; and the final cut would contain the remainder of the barium.

The first (barrier ion-containing) and the last (barium-containing) cuts from the band displacement cation exchange step will be sufficiently free of radionuclides that they can be combined with other LLW streams for final treatment. The second cut, which contains portions of the zinc barrier ion and of the Na_3EDTA solution used in developing the bands, in addition to the strontium (and a portion of the barium), could be processed further to remove the zinc and the Na_3EDTA. However, preliminary

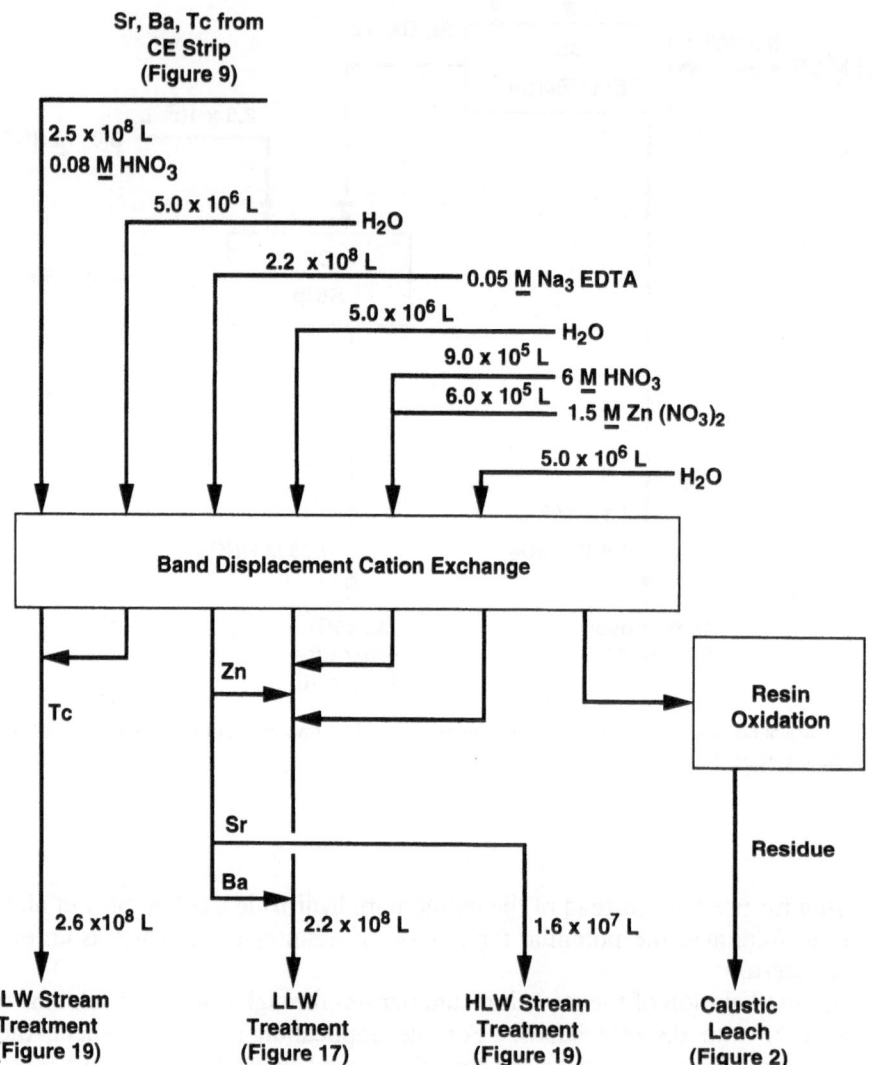

Figure 10. Separation of technetium, strontium, and barium by cation exchange.

analysis indicates such processing would not be needed to achieve the 1000-canister objective. If it is needed, it could be accomplished, after addition of nitric acid, in a second CE extraction plus stripping cycle.

Ammonium Phosphomolybdate (APM) Ion Exchange for Cesium Removal. The final radionuclide removal step currently included for treatment of the dissolved sludge solution is for removal of Cs-137 (removal of Ni-63 may be included in the future, as was discussed earlier). No completely satisfactory solvent extraction process has yet been developed for cesium removal from strong acid solution, but several potential ion exchange/sorption processes have been studied with promising results.

The choice of sorption by APM for the removal of cesium from dissolved sludge solution was based primarily on the work of Faubel and Ali (1986). These workers investigated removal of cesium from acidic solutions (some of which contained 7×10^{-3} $\underline{M}$ Fe and 9×10^{-4} $\underline{M}$ Zr) by several inorganic ion exchangers. Ammonium phosphomolybdate, $(NH_4)_3P(Mo_3O_{10})_4$, gave the best loading at nitric acid concentrations of several molar (as is the case in the CE cycle raffinate), and was chosen for use here. Faubel and Ali measured 100% retention of cesium up to a cesium loading of 50 g/kg APM in batch experiments.

In his recent review, Kolarik (1991) discusses the difficulty in eluting cesium from APM, but comments that the sorbent can be destroyed by NaOH. Thus, the example flowsheet assumes that the loaded sorbent will be introduced to the caustic leach step of the sludge dissolution process, where it will be dissolved and the components will flow to basic-side treatment processing for removal of the contained cesium along with that present in the water-soluble portions of the tank contents.

As pointed out by Faubel and Ali (1986), the use of undiluted APM in a column mode is difficult because of the microcrystalline structure of the material and because of the potential need to provide a cooling system to prevent heatspots from the radiolytic heat generation. Thus, it is assumed that the APM will be prepared or mixed with a suitable substrate that gives the required flow properties and reduces the average heat load on the column. For purposes of this example, a tenfold dilution of APM with alumina is assumed.

A depiction of this process is shown in Figure 11. A cesium DF of 10^3 in this process is assumed here; with 10% of the total cesium in this stream, an overall DF of 10^4 can still be maintained. The quantity of APM specified here assumes that the dissolved sludge solution contains 10% of the total cesium present in the waste, and that the APM is loaded to 50 g Cs/kg APM. This weight of APM equates to ~600 mole, so the outlined process will result in the addition of ~600 mole phosphorus and ~7200 mole molybdenum to the waste; these quantities are very small relative to the other components of the waste, and should have no effect on waste volumes. The ~1800 mole of ammonia that will be released during dissolution of APM in the hot caustic solution may have to be scrubbed from the offgas stream in order to meet ammonia release limits. The resulting scrub solution could be combined with other basic-side waste solutions for disposal; those solutions probably already contain considerable amounts of ammonia because most of the $\sim 10^7$ mole of fluoride in the waste (Table 1) was added to the process as ammonium fluoride solution.

Basic-Side Separations

As indicated in Figure 2, the feed to the basic-side separations processes is composed of solutions from the dissolved salt cake, from the caustic leach step, and from sludge washing operations. It will contain the bulk of the sodium, aluminum, and silicon present in the waste as well as essentially all of the waste anions. It is expected

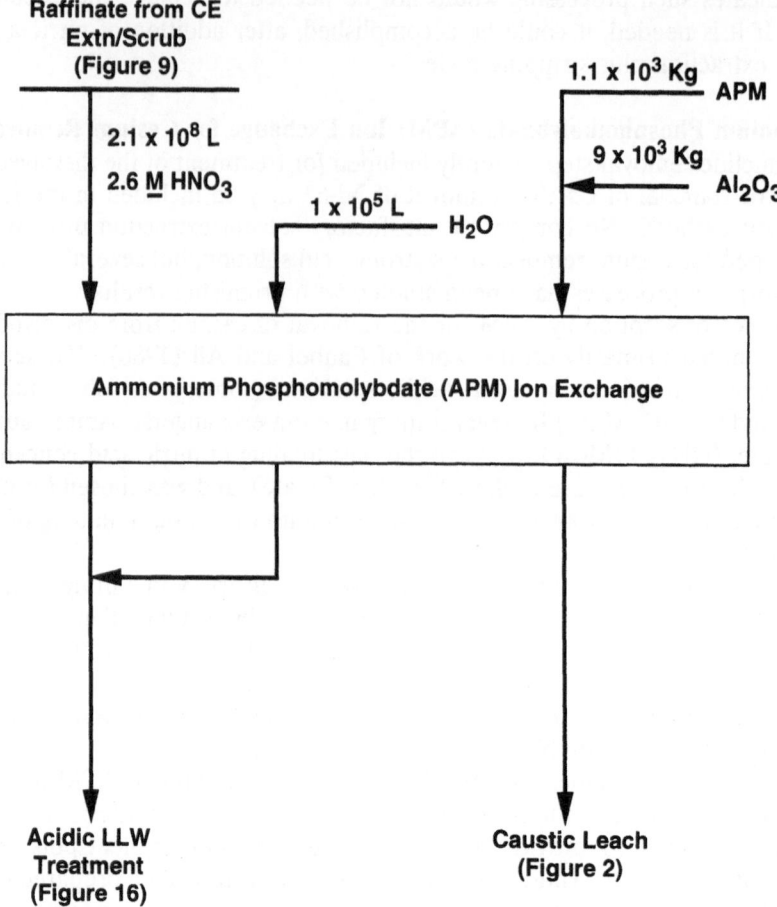

Figure 11. Ammonium phosphomolybdate ion exchange for removal of cesium from crown ether cycle raffinate.

to contain most of the Cs-137 and Tc-99 present in the waste, plus a significant fraction of the Sr-90 and possibly traces of transuranic elements (especially if organic complexants are present in sufficient concentration). This stream is also expected to contain most of the I-129 present in the waste, especially after addition of the TBP #1 solvent wash solution, which will contain extractable iodine from the dissolved sludge solution.

The initial processing steps selected for the stream in this example flowsheet are outlined in Figure 12. Following addition of a solvent wash solution and an offgas scrub solution, an evaporative concentration step is performed to decrease the volume of solution to be processed; evaporation to a sodium concentration of 5 $\underline{M}$ is assumed.

An optional ferric hydroxide scavenging precipitation step is also shown as a possible initial basic-side processing step. This is not thought to be necessary (especially in light of the relatively low required plutonium DF shown in Table 5) unless the caustic

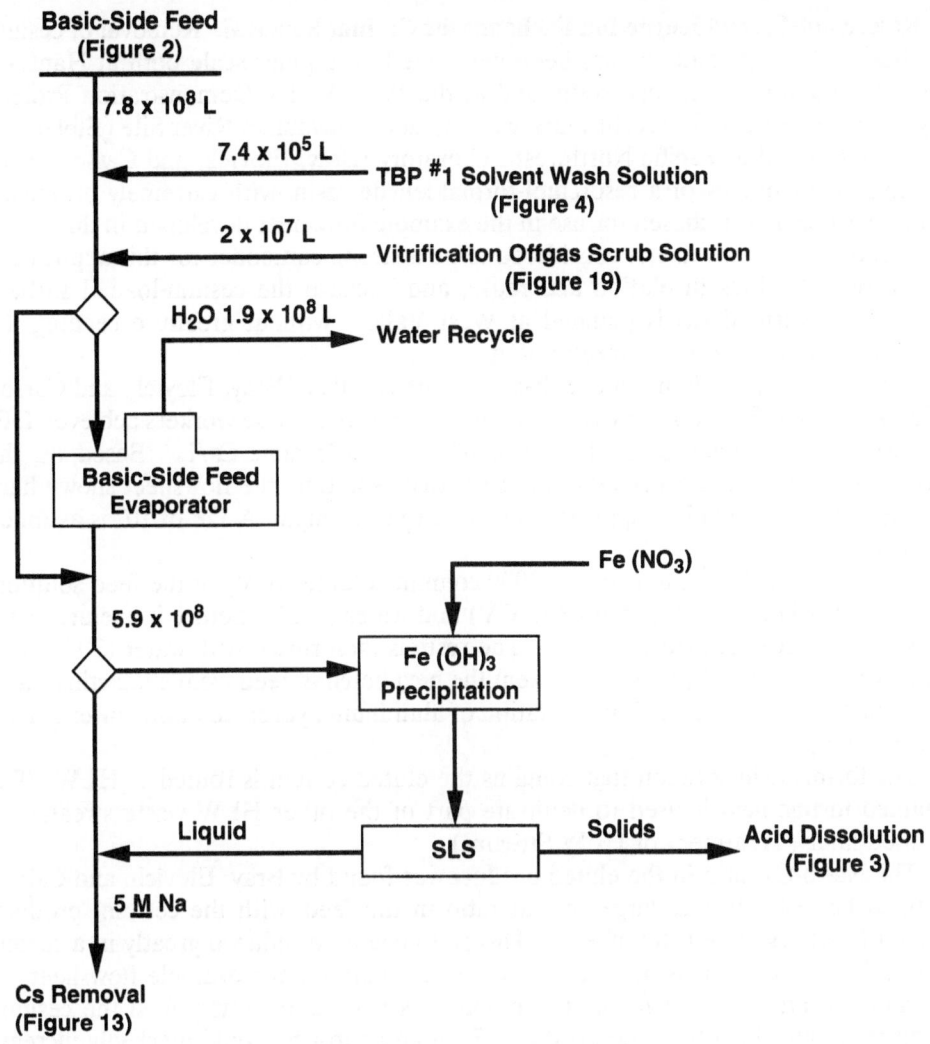

Figure 12. Preparation of feed for basic-side processing of supernate, dissolved salt cake, and caustic leach solutions.

leach step solubilizes too much plutonium from the sludge, as discussed under Sludge Washing and Dissolution. If this step is necessary, it might require that a reductant also be added to reduce Pu(V), which is the soluble species and not well scavenged by ferric hydroxide, to Pu(IV). The iron addition could come either from an added process chemical or by recycling a portion of the concentrated LLW stream from acid-side processing.

The radionuclide removal steps included in the example flowsheet for the basic-side solutions are: 1) cesium removal using a resorcinol-formaldehyde cation exchange resin, 2) strontium removal using a chelating cation exchange resin, and 3) technetium removal using an anion exchange resin. These steps are discussed further in the following sections.

Resorcinol-Formaldehyde Ion Exchange for Cesium Removal. Removal of cesium from basic-side feed solutions has been performed on a plant scale both at Hanford, using a phenolic ion exchange resin, and at the West Valley Demonstration Project, using a zeolite sorbent. In recent years, workers at the Savannah River Site (Bibler and Wallace 1987) and at Pacific Northwest Laboratory (Bray, Elovich, and Carson 1990) have conducted studies of a resorcinol-formaldehyde resin, with extremely promising results. This resin was chosen for use in the example flowsheet developed in this study. Use of zeolite sorbent was not considered to be a strong contender for this application because of difficulties in elution and reuse, and because the cesium-loaded sorbent could not be vitrified (as is planned at West Valley) without greatly exceeding the 1000-canister objective of the clean option.

A flowsheet for cesium removal based on the results of Bray, Elovich, and Carson (1990) is shown in Figure 13. By running columns in series, these workers achieved DFs $>10^4$ with cesium loadings on the resin of up to 0.05 mole Cs/L. Based on the concentration of cesium expected in the SST waste solution, the flowsheet shown here would give a cesium loading approximately one-third as high. A DF of 10^4 is assumed for this example.

After loading the cesium from ~2000 column volumes (CV) of the feed solution, the resin is washed with 2 $\underline{M}$ NaOH (3 CV) and water (6 CV) before it is eluted with 20 CV of 1 $\underline{M}$ HCOOH (formic acid). The resin is then rinsed with water (3 CV) and washed with 2 $\underline{M}$ NaOH (3 CV) to prevent the next batch of feed from contacting water (which might lead to localized precipitation of aluminum hydroxide at the lower NaOH concentration).

The formic acid solution that contains the eluted cesium is routed to HLW. The contained formic acid is used to denitrate part of the other HLW waste streams, as discussed later (Treatment of HLW Streams).

The Na-to-Cs ratio in the eluted product was found by Bray, Elovich, and Carson (1990) to be ~7 x 10^{-4} as large as that ratio in the feed, with the cesium "product" having a Na-to-Cs mole ratio of ~10. This ratio could be reduced greatly in a second cycle, if desired, but this is not thought to be necessary for the example flowsheet.

An alternative processing approach with this resin is to simply feed the cesium-loaded resin into the vitrification system. Such an approach would markedly increase resin consumption and would also become less feasible as the quantity of glass becomes smaller, because of the amount of carbon added per unit of glass.

Chelating Ion Exchange for Strontium Removal. In a recent study of the removal of strontium from basic-side wastes similar to that being considered here, Campbell and Lee (1991) chose sodium titanate as the most promising for these applications, with organic chelating resins chosen as the backup. However, the chelating resin approach was selected for this clean option example flowsheet because of the concern that direct vitrification of strontium-loaded sodium titanate along with the other radionuclide fractions might lead to an excessive number of HLW glass canisters.

In their study of chelating resins, Campbell and Lee (1991) found several to be quite effective in removing strontium from their basic-side feed solution, and to show potential column capacities in the vicinity of 2000 to 3000 bed volumes. Presumably, such capacities can also be achieved with basic-side feed solutions from the Hanford SST (and other) wastes; the achievable capacity might depend on the concentrations of calcium and barium, as well as that of strontium, because, as pointed out by Campbell and Lee (1991), removal of strontium might also require removal of these elements.

The flowsheet assumed for this strontium removal process, shown in Figure 14, is very similar to that used for cesium removal (Figure 13). After loading the strontium from ~2000 CV of the feed solution, the resin is washed with ~2 $\underline{M}$ NaOH and water

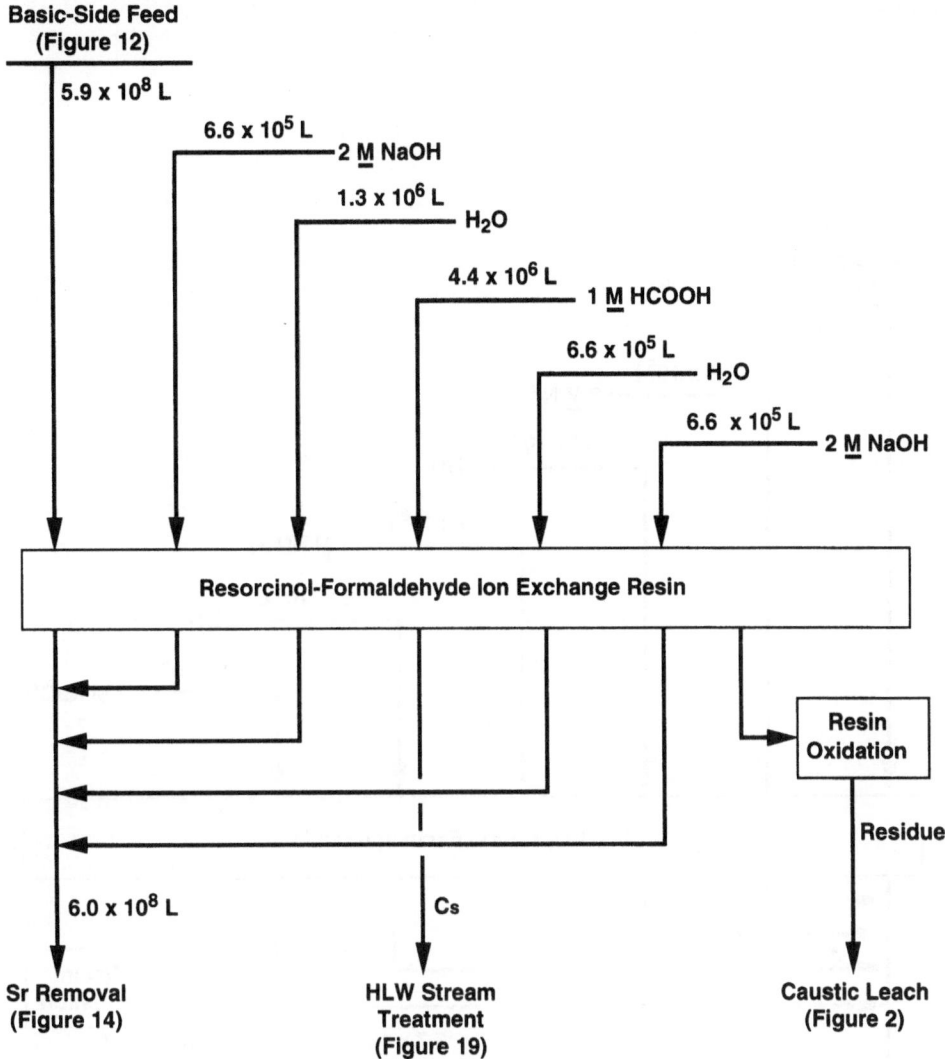

Figure 13. Resorcinol-formaldehyde ion exchange for removal of cesium from basic-side feed.

and is then eluted with ~1 $\underline{M}$ HNO$_3$. The resin is then washed with water and ~2 $\underline{M}$ NaOH to prepare it for the next loading cycle. As in the cesium removal case, the resin wash solutions are combined with the effluent from the loading cycle as feed to the next process step. A strontium DF of 10^4 is assumed in this step.

Because of the contamination of the strontium in this eluted material by calcium and barium, the acidic elution solution is routed (after evaporative concentration) to the acid-side processing sequence where separation from these contaminants is achieved. The eluted strontium solution is assumed to be routed to the acid dissolution step (Figure 3), where its contained nitric acid can be put to good use in dissolving the sludge.

The choice of the chelating resin approach over the sodium titanate sorption approach for removing strontium from the basic-side feed in this example flowsheet should perhaps be reexamined in future studies. The total amount of titanate required to be added to achieve the required DF will not be known until the fraction of the

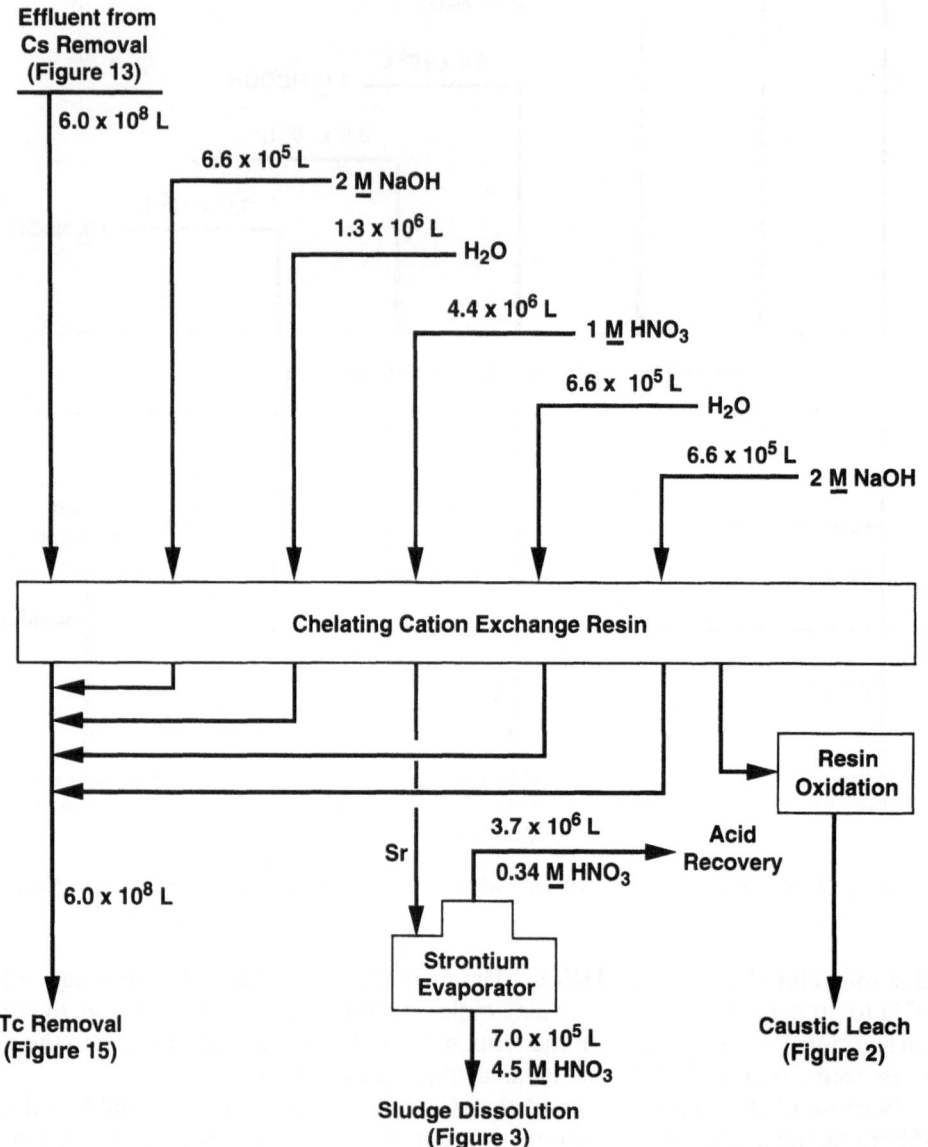

Figure 14. Chelating cation exchange for removal of strontium from resorcinol-formaldehyde process effluent.

strontium that is present in this solution has been determined. It may also be possible to avoid vitrification of the added titanium by dissolving the strontium-loaded titanate in acid and separating the strontium from the titanium in the acid-side CE cycle. Also, the concern over the impact on vitrified HLW quantity would disappear if some other disposition of separated strontium (e.g., special storage) were selected.

Anion Exchange Removal of Technetium. Anion exchange processes have been used to recover technetium from basic-side Hanford tank wastes both on a laboratory scale (Buckingham et al. 1967; Roberts, Smith, and Wheelwright 1963; Schulz 1980) and on a plant scale (Beard and Caudill 1964; General Electric 1964). Because of this experience, such a process was included in the flowsheet of this study.

In this process, the pertechnetate ion is removed from the basic-side solution by sorption on a strong-base anion exchange resin. After the resin bed is loaded, it is washed first with dilute sodium/hydroxide to remove residual feed solution and then with dilute nitric acid to ensure conversion of the resin to the nitrate form before the elution step is done. The technetium is eluted with 6 $\underline{M}$ HNO_3, and the bed is prepared for the next loading cycle by displacing the acid first with water and then with dilute NaOH.

The technetium product stream is concentrated, and the distilled nitric acid is condensed and reused in the next elution cycle. A small amount of technetium may be volatilized during concentration, but it accompanies the acid to the next elution cycle and thus is not lost from the system.

A flowsheet for this technetium removal step is shown in Figure 15. It is based on an early flowsheet (Beard and Caudill 1964) used to recover technetium from Hanford tank wastes, but has been modified here to include a higher elution volume in order to ensure higher removal of technetium from the waste (the early work was aimed at recovering most of the technetium for beneficial use, rather than at attaining high removal from the waste). Based on the results of Schulz (1980), loading and elution volumes of ~50 CV of feed and ~20 CV of 6 $\underline{M}$ HNO_3 eluant are used in this flow-sheet, and a technetium DF of 10^2 is assumed in this step.

Because technetium is sorbed more strongly from hydroxide solutions than from alkaline nitrate solutions, a more efficient technetium removal process could result if it were done after, instead of before, the nitrate destruction step, as in the current flowsheet (Figure 1). Volatilization of technetium during calcination, the currently indicated nitrate destruction step, could pose a problem to be addressed in that case, however.

Treatment of LLW Streams Resulting from Separations

The LLW streams resulting from the separations steps described above are treated further before they are finally disposed of. These treatments are discussed in this section, first for the LLW streams from the acid-side processing steps and then for the basic-side processing LLW stream. The recovery for recycle of NaOH from the treated LLW will also be described.

LLW Streams from Acid-Side Separations. A variety of LLW streams occur from the acid-side separations processes described earlier. The primary stream is the highly acidic, decontaminated, dissolved sludge solution. The ion exchange process to separate americium from the bulk of the lanthanides gives both an acid-containing waste solution and a near-neutral waste solution. A near-neutral waste solution also results from the

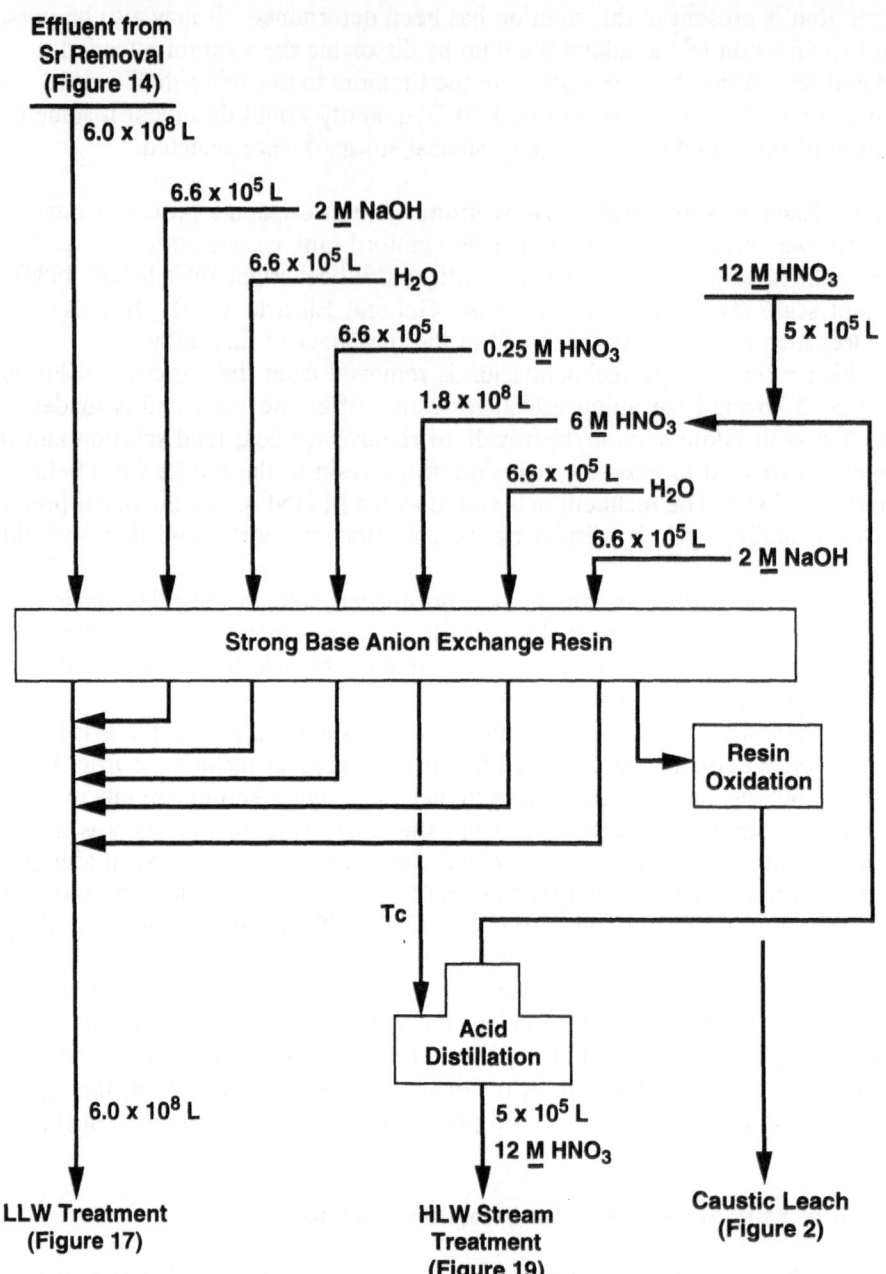

Figure 15. Strong-base anion exchange for removal of technetium from chelating cation exchange process effluent.

ion exchange process used to separate strontium from barium. Basic LLW solutions are also present as a result of the solvent washing operations in the TBP and CMPO solvent extraction processes.

A common feature of several of these LLW streams is that they contain organic complexants (e.g., oxalate, DTPA, EDTA) that were added in the dissolution and the separations steps. Destruction of these complexants could be performed at this point of the process (e.g., by reaction with the nitric acid and/or with added hydrogen peroxide) or, alternatively, the destruction could be performed after these acid-side processing LLW streams are combined with the basic-side processing LLW streams. The latter approach to ultimate destruction of organic complexants was chosen for this example flowsheet, with some preliminary destruction likely occurring during acid recovery.

There are also choices regarding the approach to be followed for disposition of the HNO_3 contained in these LLW streams. The HNO_3 could be converted to $NaNO_3$ with NaOH, or (most of) it could be recovered for reuse. If it were converted to $NaNO_3$ with new NaOH, an increase in the final volume of treated LLW would result; however, such an increase would not result if recycled NaOH were used. If most of the HNO_3 is recovered for reuse, the quantity of NaOH required for conversion to $NaNO_3$ is minimized. The approach selected for the example flowsheet includes HNO_3 recovery (from selected streams); this approach is shown in Figure 16.

Figure 16 identifies the two acid-side separations LLW streams from which HNO_3 is to be recovered. The first step in the recovery process is an evaporative step to reduce the volume of solution to be treated; some of the HNO_3 is recovered by distillation in this process. Aluminum nitrate is assumed to be added to the evaporator in order to minimize corrosion from the fluoride complexant that was used in sludge dissolution (Figure 3). The organic complexants that are present are expected to be partially destroyed during this evaporation step.

The bulk of the HNO_3 recovery occurs from recovery of the NO and NO_2 (NO_x) that are evolved when the concentrated waste solution is denitrated by reaction with sucrose. The NO_x streams from this denitration operation will be combined with others, and treated to convert NO_x back to HNO_3 (see Recycling of Nitric Acid and Water).

LLW Stream from Basic-Side Separations. The only LLW stream resulting from the basic-side separations processes described earlier is the raffinate from the strong-base anion exchange process used for technetium removal (Figure 15). This is the case because the solutions used to wash and regenerate the ion exchange resins are of such small volume that they could be added to the mainline dissolved salt cake solution without markedly increasing its volume.

This LLW stream contains large quantities of nitrate and nitrite, and perhaps some organic complexants, from the SST inventory. In the example flowsheet, it is planned that the waste be calcined to destroy these constituents before the LLW is finally disposed of; this is done in order to increase the acceptability of the LLW. The LLW streams from the acid-side separations steps should also be calcined; it appears to be reasonable to calcine all the LLW streams in one combined operation.

Figure 17 depicts this approach. Four basic or near-neutral LLW solutions from the acid-side processing operations are added to the basic-side processing LLW stream and are assumed to be concentrated by evaporation (to decrease the quantity of water to be fed to the calciner); the evaporated water is condensed for reuse. The denitrated, acid-side LLW solution is added to the evaporator bottoms solution, and the resulting (basic) mixture is calcined.

Figure 17 shows removal of iodine from the calciner offgas stream. This is based on the assumption that the iodine species fed to the calciner will volatilize, most likely as I_2 or as organic iodides. Removal of such volatile species by sorption on silver-impregnated sorbents has been studied extensively, and is assumed for this example flowsheet. However, great uncertainty exists regarding this iodine removal operation. Little is known regarding such important factors as 1) the quantity of iodine that is actually present, 2) the nature of the iodine species present, and 3) the degrees of volatilization of these species during calcination. Because of these uncertainties, a low iodine DF of 10 was assumed for this example.

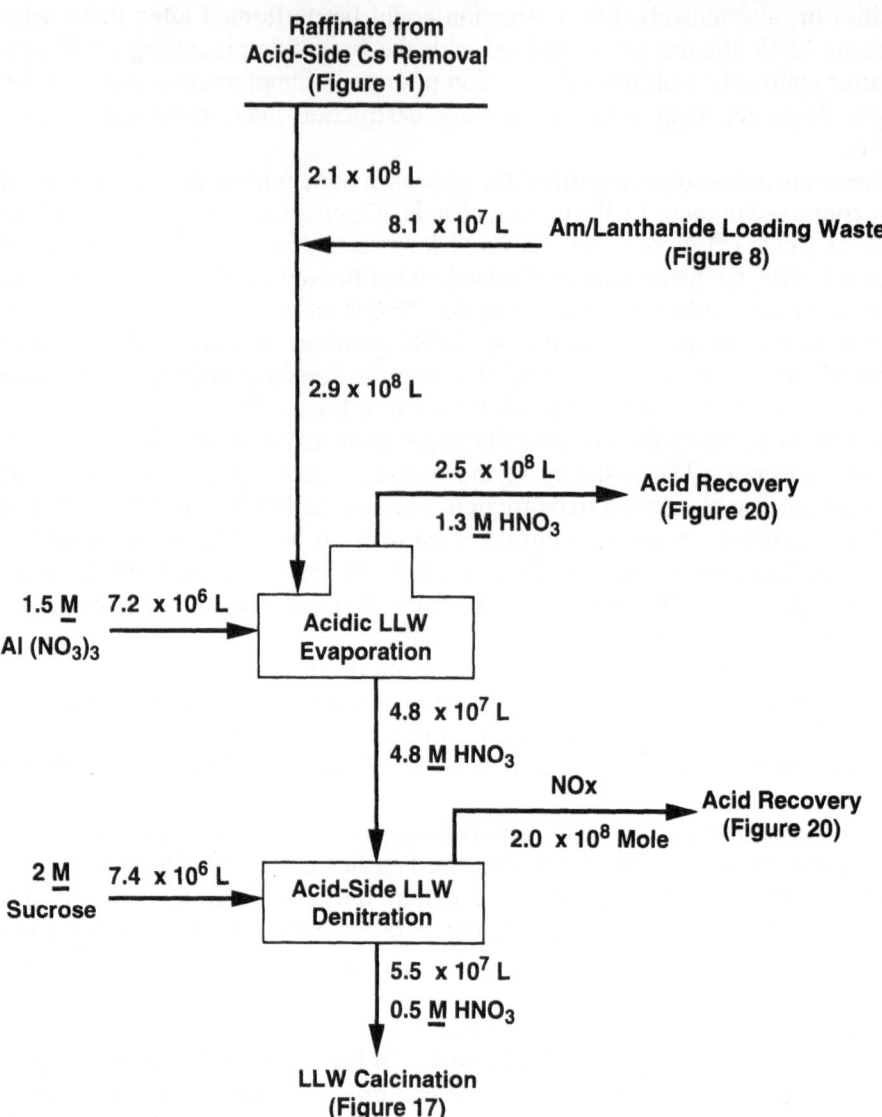

Figure 16. Initial treatment of acidic low-level waste streams.

All of the NO_x produced in the calcination process is assumed in this example to be destroyed (converted to N_2 and H_2O) by a catalyzed reaction with NH_3. Alternatively, nitric acid could be recovered from the NO_x stream to decrease the amount of NO_x to be destroyed. The quantity of NO_x shown in Figure 17 is based on the assumption that 10% of the nitrogen in $NaNO_3$ and $NaNO_2$ is evolved as NO_x (with the remainder being evolved as N_2).

Finally, water is condensed from the calcination offgas. This water is mixed back with the calcined material to give a slurry of insoluble metal hydroxides in a NaOH solution.

Sodium Hydroxide Reuse and Disposal. The caustic slurry produced in the calcination operation is a source of sodium hydroxide that can be recovered for process

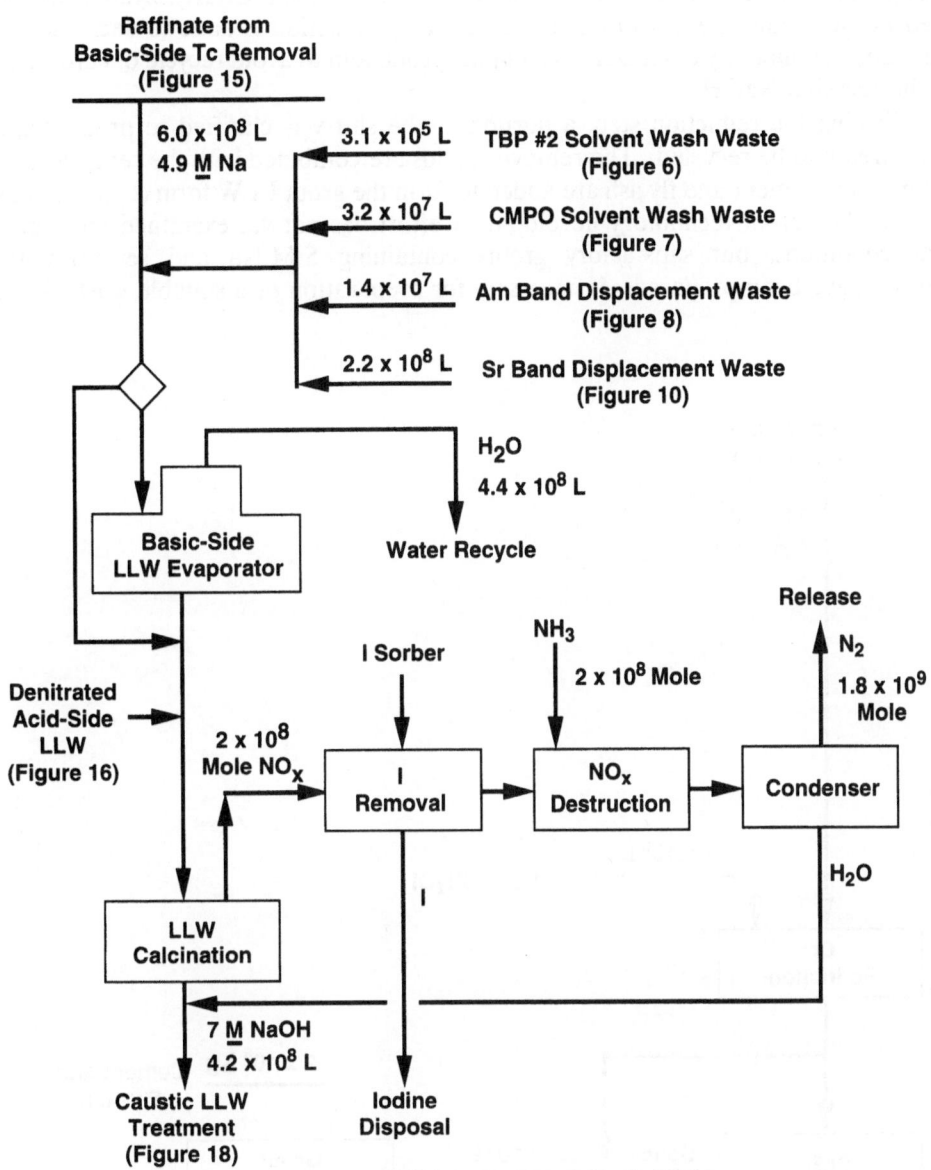

Figure 17. Treatment of low-level waste streams.

uses, thus minimizing the increase in LLW disposal volume that results from the separations processing described in the preceding sections. All of the NaOH will ultimately be disposed of in the (grout) LLW form but, during processing operations, a portion of the NaOH solution can be reused.

Figure 18 outlines the steps anticipated for these reuse and disposal operations. The first step involves reduction of soluble Cr(VI) to insoluble Cr(III); hydroxylamine is used as the reducing agent in this example. This reduction is done to decrease the leachability and mobility of chromium from the grout, and to avoid recycle of chromium with the recycled NaOH.

Following the reduction step, a portion of the slurry is clarified to produce the NaOH stream to be recycled. The removed solids are combined with the remainder of the slurry, and cement and fly ash are added to form the grout LLW form (that contains 5 $\underline{M}$ Na). The grout technology development effort has not yet examined these particular conditions, but satisfactory grouts containing 5 $\underline{M}$ Na and several-molar hydroxide have been prepared. If necessary for preparation of a suitable waste form,

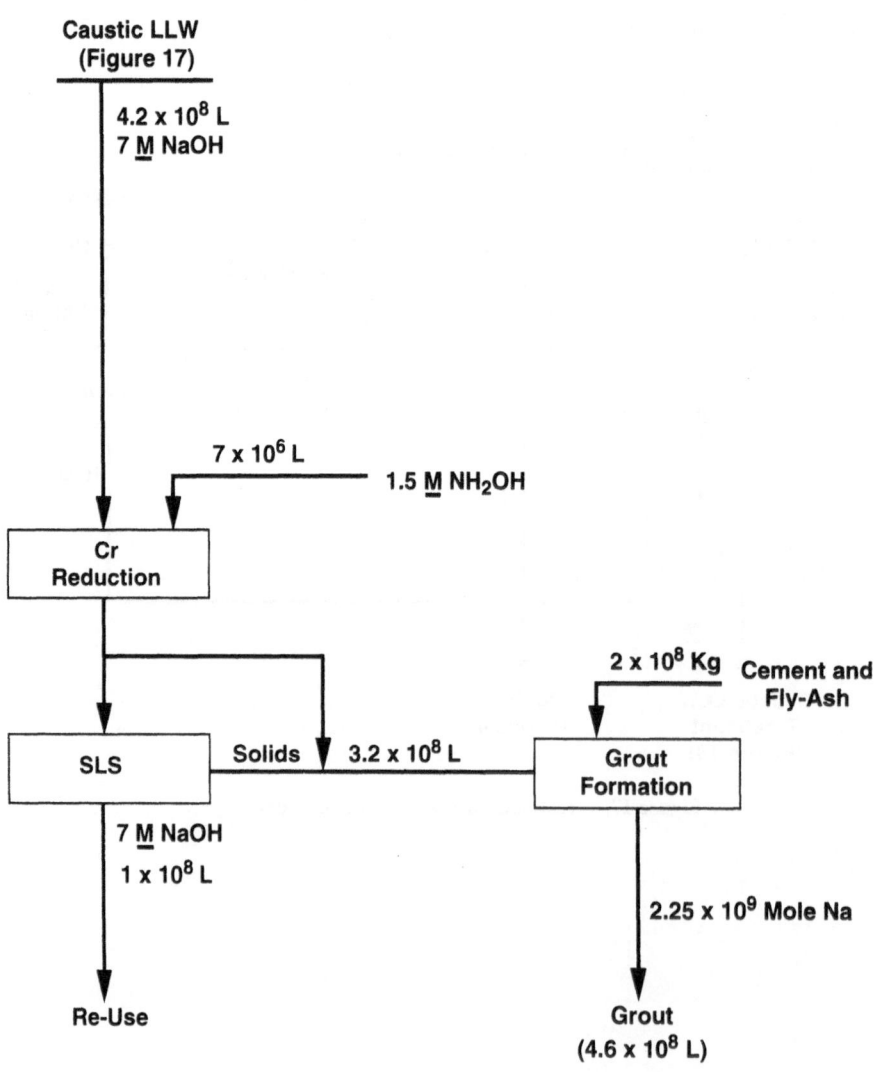

Figure 18. Caustic treatment and grout formation.

a portion of the NaOH could presumably be converted to Na_2SO_4 (or other salt) before grouting without affecting the 5 $\underline{M}$ Na in grout assumption of this flowsheet.

The recycled NaOH stream is expected to contain other soluble salts. This should not present significant problems because ~5 times as much NaOH goes to grout as is recycled, which should prevent appreciable build-up of any contaminant in the recycled NaOH.

Treatment of HLW Streams

The process steps discussed in the Acid-side and Basic-side Separations sections result in the generation of six acidic streams containing (partially) separated radionuclides that are to be disposed of in HLW glass. One of these streams is a formic acid stream containing the cesium. The other five contain nitric acid; these are 1) the plutonium, neptunium, and thorium stream from the TBP cycle; 2) the americium and heavy lanthanide stream from the CMPO cycle plus americium/lanthanide ion exchange; 3) the technetium stream from the CE cycle plus strontium/barium ion exchange; 4) the strontium stream from the CE cycle plus strontium/barium ion exchange; and 5) the technetium stream from the anion exchange cycle.

These streams could presumably be sent to the HLW glass melter without further treatment, but that is not likely to be desirable because of their diluteness. Thus, the example flowsheet includes provisions to concentrate and partially denitrate these streams and to recover the HNO_3 and NO_x resulting from these operations. The flowsheet shown in Figure 19 outlines these steps, and also includes for completeness the materials given off during operation of the melter itself.

The HNO_3-containing streams are first concentrated by evaporation; this concentration step is not carried to a very small volume in order to avoid the possibility of precipitation of strontium nitrate in the evaporator. The quantity of organic complexing agents (DTPA and EDTA) present in the americium, samarium, and strontium product solutions might need to be reduced to meet the 1000-canister objective. If sufficient complexant destruction does not occur in the simple conditions outlined in Figure 19, then hydrogen peroxide (plus catalyst) could be added to the evaporator to increase the extent of destruction. The evaporator bottoms are then denitrated by reaction with sugar, as well as with the formic acid in the cesium solution.

The concentrated and denitrated solution, which contains only ~3% of the H_2O and ~10% of the HNO_3 that was present in the feed streams, then enters the glass melter. The acidic melter offgas is scrubbed with NaOH solution to remove the contained H_2O and HNO_3 (and a portion of the NO_x evolved from the metal nitrates); the scrubber will also remove the portions of the cesium and technetium that volatilize from the melter. This scrubber solution is indicated to be routed to the basic-side processing feed (Figure 12) to prevent loss of the cesium and technetium, although it could perhaps be handled otherwise in the vitrification facility itself.

Recycling of Nitric Acid and Water

In this example flowsheet, where the wastes contain large quantities of sodium and where recycled NaOH is used to neutralize excess acids, recycling of HNO_3 is not needed in order to minimize the increase in LLW disposal volume resulting from processing the waste. However, recycling of HNO_3 is included in the flowsheet to minimize the release of nitrate destruction products to the atmosphere, to decrease the load on the calcination and NO_x destruction operations, and to increase the validity of this flowsheet for wastes having lower ratios of sodium salts to insoluble hydroxide

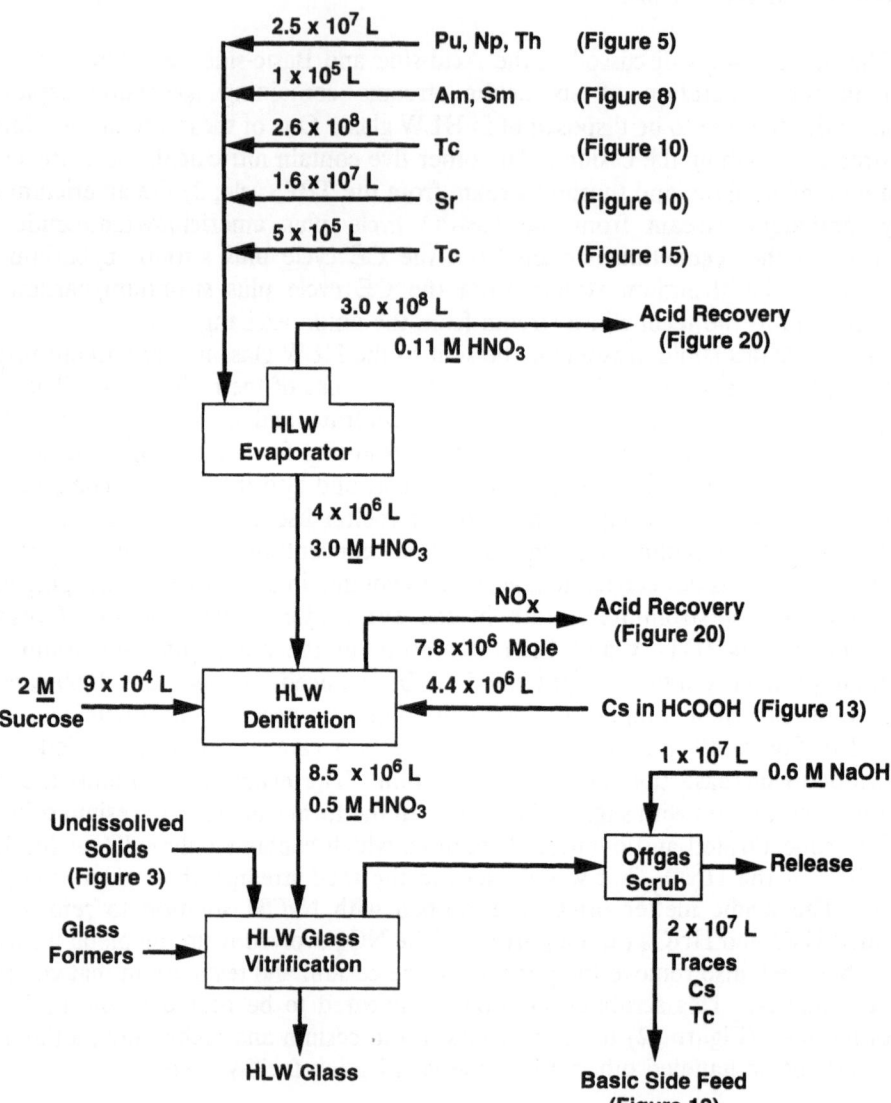

Figure 19. Treatment of high-level waste streams.

sludges. Similarly, recycling of H_2O is included to minimize the quantity of it that must be disposed of in LLW, or discharged to the atmosphere.

Figure 20 shows a flow diagram for the recovery of HNO_3 from the overheads streams of acidic evaporators and from the gaseous NO_x streams resulting from denitration of two liquid streams and from calcination of the purified uranium product. These streams are contacted in a suitable absorber to convert the NO_x to HNO_3; oxygen (O_2) and hydrogen peroxide (H_2O_2) are added to aid in this conversion. The gaseous effluent from this absorber is discharged to the atmosphere; presumably the absorber efficiency will be high enough that the effluent will meet NO_x discharge limits; otherwise, a residual NO_x destruction step would be required. The liquid HNO_3 stream from the NO_x absorber feeds a fractional distillation column where the HNO_3 is fractionated into a concentrated (12 $\underline{M}$) bottoms solutions and a dilute (~0.005 $\underline{M}$) overheads stream, both of which are recycled to the process.

The quantity of HNO_3 recovered from the acid fractionator as 12 $\underline{M}$ HNO_3 (5.8×10^8 mole) is only $\sim80\%$ of the HNO_3 inputs to the sludge dissolution and the separations steps. Thus, a net system addition of $\sim1.5 \times 10^8$ mole HNO_3 will be required to process the SST waste according to the example flowsheet presented here. This added HNO_3 is effectively converted to $NaNO_3$ by part of the NaOH present in the SST waste, giving a small increase in the total nitrate content of the waste to be calcined.

A detailed water balance has not been performed for the example flowsheet. However, because the SST wastes currently contain $\sim4 \times 10^7$ L H_2O, and the grout LLW form resulting from processing these wastes by the example flowsheet contains 3.1 $\times 10^8$ L H_2O, an overall net input of $\sim2.7 \times 10^8$ L H_2O is required to process the SST wastes. Water recycle can be employed to minimize water additions in excess of the required amount. In this regard, the volume of water used in the steps outlined in Figure 2 (waste retrieval and dissolution, caustic leach, and solids washing) is noted to be the same as the volume of water present in basic evaporator overheads (Figures 12 and 17). It is also noted that the total volume of water required in the dissolution (Figure 3) and strontium stripping (Figure 9) operations is the same as the total volume present in the slightly acidic evaporator and fractionator overheads (Figures 6 and 20). Thus, reasonable recycling routes do exist for the recovered condensates.

It is possible that some excess water will require disposal during processing, because of variabilities in water inputs and disposal over time. It is assumed for this example that any excess water will be disposed of by evaporation to the atmosphere rather than by incorporation in grout.

RESULTS OF FLOWSHEET IMPLEMENTATION

Results of implementing the example flowsheet can now be estimated in three important areas: 1) the quantity of HLW, 2) the quantity of LLW, and 3) the radionuclide content of the LLW. These estimates are discussed here.

Quantity of HLW Glass

The minimum quantity of glass required to contain each radionuclide separated from the SST waste can be calculated from the data presented in the Study Bases section; results of such calculations are summarized in Table 7 (along with those for the DST plus SST wastes). The total of these separate considerations for both wastes is ~1100 canisters; it is highly likely that fewer canisters will suffice because of component compatibilities. For example, strontium and technetium would not behave similarly in glass, so their individual quantity requirements should not be additive. Thus, it is highly

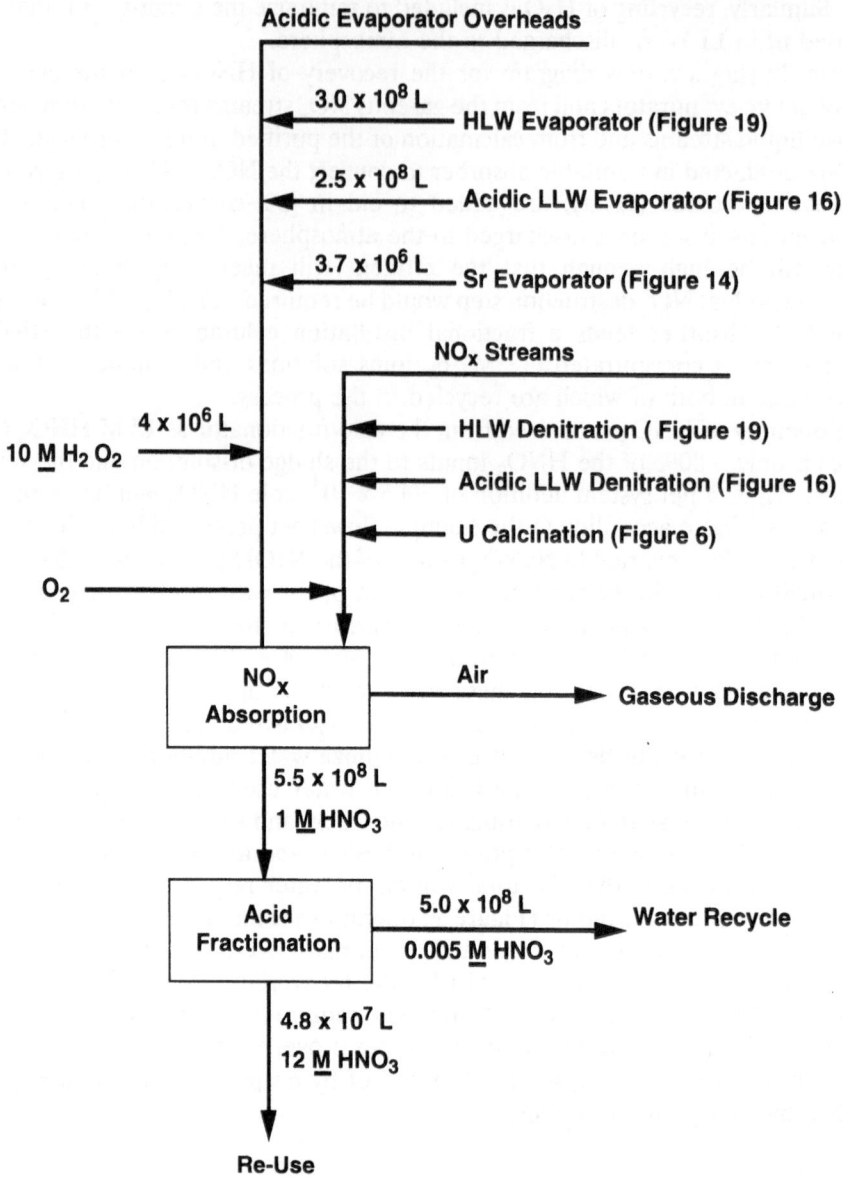

Acidic Evaporator Overheads

3.0×10^8 L — HLW Evaporator (Figure 19)

2.5×10^8 L — Acidic LLW Evaporator (Figure 16)

3.7×10^6 L — Sr Evaporator (Figure 14)

NO_x Streams

4×10^6 L

10 $\underline{M}$ H_2O_2

HLW Denitration (Figure 19)

Acidic LLW Denitration (Figure 16)

U Calcination (Figure 6)

O_2

NO_x Absorption

Air → Gaseous Discharge

5.5×10^8 L

1 $\underline{M}$ HNO_3

Acid Fractionation

5.0×10^8 L — Water Recycle

0.005 $\underline{M}$ HNO_3

4.8×10^7 L

12 $\underline{M}$ HNO_3

Re-Use

Figure 20. Nitric acid recovery.

likely that the total number of HLW glass canisters required for the separated radionuclides will be below the 1000-canister goal of the clean option, which is a 12-fold reduction in the number indicated in the current reference case.

However, as has been discussed in earlier sections, the number of HLW glass canisters resulting from implementation of the clean option could well be defined by the completeness of dissolution of the sludge components rather than by the number of canisters required to accommodate the radionuclides. Unless the undissolved material (or a blend of the undissolved material with other process waste streams) meets Class A LLW criteria, it must be disposed of in the HLW. In order to achieve the 1000-canister objective, the undissolved material must contain < ~2% of the aluminum,

Table 7. Number of HLW canisters required for radionuclides separated from wastes.

| | Number of HLW Glass Canisters Required[1] | |
Waste Component	SST Wastes Only	DST + SST Wastes
Cesium	$<1 \times 10^{1(2)}$	$<1 \times 10^{1(2)}$
Heavy Lanthanides	$<5 \times 10^1$	$<5 \times 10^1$
Strontium	5×10^2	5×10^2
Technetium	1×10^2	5×10^2
Thorium	1×10^2	1×10^2
Transuranics	$<5 \times 10^1$	$<5 \times 10^1$

[1]Considering each component separately; based on 1650 kg glass/canister, the quantities present in the wastes, and the allowed concentrations in glass (Table 5).

[2]Heat loading considerations would require a higher number of canisters.

$< \sim 1\%$ of the chromium, and $< \sim 0.1\%$ of the sodium present in SST wastes (Table 3). Much experimental testing must be done with various Hanford waste sludges before the probability of achieving the 1000-canister objective can be assessed. Results of the preliminary testing conducted to date are quite encouraging, as was discussed under Sludge Washing and Dissolution.

While the bases of this study include the current HWVP bases for feed and glass composition, it should be pointed out that decreased HLW glass volumes could result from changes in the liquid-fed ceramic melter design (as well as from the changes in process chemistry that are the thrust of this paper). A melter designed to operate at a higher temperature could produce glass with a higher waste loading, and incorporation of a bottom drain for discharge of accumulated insolubles from the melter could lead to marked reductions in the canister requirement.

LLW Grout Quantity Considerations

This section first presents estimates of the volume of LLW grout produced by processing with the example flowsheet both a) SST wastes alone and b) combined DST and SST wastes, and compares the volume for the combined waste case with that of a recent reference case estimate. The effect (actually, the lack of an effect) of process chemical additions on the LLW volume is then discussed.

LLW Volume Comparison. The volume of LLW grout resulting from processing the SST wastes is 4.6×10^8 L (4.6×10^5 m^3), as shown in Figure 18. This volume is defined almost exclusively by the chemicals that are present in the existing waste rather than by the chemicals used in processing the wastes, as will be discussed in the next section. This volume is based on a sodium concentration in the grout of 5 $\underline{M}$. It should also be mentioned again here that compositional envelopes of suitable grouts have not yet been defined.

The quantity of sodium in the DST wastes is ~36% of that in the SST wastes. Thus, the volume of grout resulting from processing both DST and SST wastes should be ~36% higher than the volume resulting from processing only SST wastes, or ~6.3 x 10^5 m^3. This volume is less than half of that indicated in the cited reference case; it would require 121 grout vaults of the size currently envisioned (each holding 5200 m^3 of grout).

This lower LLW volume that results from use of the example clean option flowsheet is not unique to that flowsheet. It can be achieved by any flowsheet that uses the philosophy of recovering and recycling water and key chemicals (especially NaOH) added during processing of the waste.

Effect of Process Chemical Additions on LLW Volume. The processes identified in the Example Flowsheet Description involve the use of many process chemicals (in addition to nitric acid and water) to accomplish the desired results. However, only a portion of the components of those chemicals will end up in the final LLW waste form, because of conversion to volatile compounds (water and gases) before the waste is finally disposed of. Furthermore, the only nonvolatile components whose additions are likely to have any impact at all on the final LLW grout volume are those that increase the quantity of a grout volume-controlling component, which appears to be sodium in mixtures such as the components of Hanford tank wastes.

Numerous process chemicals used in the example flowsheet will be converted to volatile compounds, and thus will not be present in the LLW. Organic chemicals that fall into this category are organic complexing agents (e.g., oxalate, DTPA, and EDTA) used to achieve sludge dissolution, to prevent extraction of unwanted components, and to facilitate some special separations, and sugar and formic acid used in denitration reactions. Inorganic chemicals that behave in this manner are hydroxylamine (NH$_2$OH) and hydrogen peroxide (H$_2$O$_2$).

Numerous other process chemicals will be converted to materials that will be included in the final LLW waste form. Table 8 lists these components, compares the added quantities with those that are already present in the SST wastes (Table 1), and compares the mole ratios (relative to the initial sodium content) of the initial SST waste and the LLW after processing. With sodium being the grout volume-controlling component of these wastes, these data indicate that the LLW grout resulting from the processing steps of the example flowsheet would have a volume only 0.4% greater than if the tank wastes had been grouted without treatment.

The only large addition to the LLW, relative to the amount initially present in the SST waste, is phosphate. Phosphorus is added to the waste intentionally (e.g., HEDPA and APM) and as a result of both dissolution and entrainment of phosphorus-containing extractants (e.g., TBP and CMPO) in the aqueous phases leaving the solvent extraction contractors. The dissolution and entrainment additions are much larger than the intentional additions; assuming TBP and CMPO solubilities of 0.1 g/L and entrainment amounting to 10^{-3} L of solvent per liter of aqueous solution, the example flowsheet would result in the addition of ~8 x 10^7 mole phosphorus to the waste. Assuming that all of this phosphorus is converted to phosphate in waste evaporation and/or calcination steps, the amount added to the waste by the indicated processing would be ~90% of that already present in the waste (Table 8). This is not surprising since most of the waste resulted from processes employing phosphorus-containing extractants. However, even this large increase in the phosphate content of the waste should have no effect on the volume of LLW grout resulting from SST waste processing, because the amount of phosphate will still be small relative to sodium (0.076 mole per mole).

Table 8. Additions to LLW caused by processing SST wastes.

Component	Initially Present in SST, Mole	Added to LLW During Processing, Mole	Added, % of Initial	Mole per Mole Na in SST	
				In SST	In LLW After Processing
Na	2.25×10^9	$1.0 \times 10^{7(1)}$	0.4	1.000	1.004
PO_4	9.20×10^7	8×10^7	~90	0.041	0.076
Al	9.05×10^7	1.1×10^7	12	0.040	0.045
F	4.24×10^7	4.5×10^6	11	0.019	0.021
SO_4	1.72×10^7	1.4×10^6	8	0.0076	0.0083
Zn	--	1.4×10^6	--	--	0.0006

[1]An additional 7.5×10^8 mole Na is used in processing, but does not add to the LLW because recycle NaOH is used.

The major use of nonvolatile materials for the flowsheet operations is in compounds of sodium (e.g., NaOH, Na_2CO_3, $NaNO_2$, Na_3DTPA, Na_3EDTA). Most of this usage involves NaOH in the caustic leach step, in partial neutralization and adjustment of DTPA and EDTA solutions to the pH values needed for proper separations, and in ion exchange column washes. Because recycle NaOH is used for these operations, no sodium addition to the waste will result. Sodium used in the Na_2CO_3 solutions used for solvent washing and in the $NaNO_2$ solution used for neptunium oxidation will be added to the waste; this quantity is listed in Table 8.

Sulfate is added to the LLW as a residue from the oxidation of some of the spent ion exchange resins used in the processing. The resins commonly used for the band displacement cation exchange separations of americium and of strontium are sulfonated polystyrenes, which will yield sulfate when oxidized. This is not the case with the resins used for the basic-side removals of cesium, barium, and technetium. The added sulfate value listed in Table 8 assumes that the contribution from the resin used for strontium separation is the same as the contribution from the resin used for americium separation, which was estimated from the resin usage estimate given by Wheelwright et al. (1974) for a process involving recovery of curium from power reactor fuel wastes. Thus, the sulfate value listed here is likely a conservative (high) one.

Fluoride is used in the example flowsheet to assist in sludge dissolution, and aluminum is added later to minimize fluoride-enhanced corrosion during acidic waste evaporation and denitration. Neither addition is large relative to the initial SST contents, and the total quantities are small relative to sodium. Zinc and molybdenum are not listed as constituents of the SST waste, so the percentage increases by use of the flowsheet cannot be given. However, the added zinc and molybdenum are small fractions of other components.

Radionuclide Content of LLW Grout

Table 9 summarizes the DFs assumed to result from the separations steps, lists the quantities and concentrations of radionuclides present in the LLW, and compares the

Table 9. Radionuclides in clean option LLW from SST wastes.

Radionuclide	Assumed Treatment DF	Quantity in LLW, Ci	Concentration in Average SST LLW, Ci/m^3	Fraction of Class A Limit
Class A Listed				
C-14	1	1.4×10^2	3×10^{-4}	4×10^{-4}
Ni-63	1	3×10^5	6×10^{-1}	2×10^{-1}
Sr-90	10^4	4.5×10^3	1×10^{-2}	2×10^{-1}
Tc-99	10^2	9×10^1	2×10^{-4}	7×10^{-4}
I-129	10^1	2.4	5×10^{-6}	6×10^{-4}
Cs-137	10^4	9.5×10^2	2×10^{-3}	2×10^{-3}
Np-237	10^2	3.1×10^{-1}	7×10^{-7}	4×10^{-5}
Pu-239	10^4	2.7	6×10^{-6}	4×10^{-4}
Am-241	10^4	3.6	8×10^{-6}	5×10^{-4}
Other				
Se-79	1	1×10^1	3×10^{-5}	
Zr-93	1	4×10^3	9×10^{-3}	
Pd-107	1	3×10^1	7×10^{-5}	
Sn-126	1	4×10^2	1×10^{-3}	
Sm-151	10^4	4×10^1	1×10^{-4}	
Th-232	10^2	1×10^{-2}	3×10^{-8}	
U-238	10^4	4×10^{-2}	9×10^{-8}	
		3.1×10^5	6.2×10^{-1}	4×10^{-1}

radionuclide concentrations in the average SST LLW with the Class A limits. The radionuclides whose concentrations are closest to the Class A limits are Ni-63 and Sr-90, each of which is present at ~20% of the limit, on an overall average basis. As has been discussed earlier, if it is required that each increment of the LLW grout meets the Class A standards, then higher DFs than those indicated for Ni-63 and Sr-90 will be required when the contents of some tanks are processed, because of expected concentration variabilities among the tanks. Alternatively, the time of reaching the Class A limit could be established as being a few hundred years in the future to take advantage of the decay of these radionuclides (the half-lives of Ni-63 and Sr-90 are 100 and 29 years, respectively).

In the current flowsheet, where nickel removal is not included, Ni-63 contributes nearly all of the 0.31 MCi total radionuclide content of the SST LLW content. This total content is a factor of ~200 lower than in the untreated SST wastes. Addition of a 99% efficient nickel removal step would decrease the total radionuclide content to 0.013 MCi, which is lower by another factor of ~20.

If the total DST wastes were also processed by the same flowsheet, an additional 0.025 MCi of total radioactivity (~80% of it from Ni-63) would be added to the LLW inventory (and the overall average DST LLW would contain ~35% of the Sr-90 allowed by the Class A limit). Thus, the LLW from processing the combined SST and DST wastes by the example clean option flowsheet (no Ni-63 removal) would contain ~0.34 MCi. This LLW radionuclide content is a factor of ~50 lower than the ~18 MCi value (with daughter activity subtracted) given in the reference case.

ACKNOWLEDGMENTS

The assistance of the other persons involved in the early considerations of the clean option concept is gratefully acknowledged. They are Jerry L. Straalsund, Eddie G. Baker, Evan O. Jones, and William L. Kuhn of Pacific Northwest Laboratory and James J. Holmes of Westinghouse Hanford Company. Dr. Straalsund deserves special recognition as the driving force for the development of this concept.

Special thanks are due to David O. Campbell for several valuable discussions and for identification of some technical errors in the initial draft of this paper, and to Wallace W. Schulz who questioned the need for some steps in the initial draft flowsheet and thus contributed to some flowsheet simplification. Valuable discussions with George F. Vandegrift and Gregg J. Lumetta are also greatly appreciated. The assistance received from the other national technical experts identified below who contributed their time to reviewing the draft of this paper, especially those who also attended the review meeting at PNL in November 1992, is also gratefully acknowledged. Helpful discussions were also held with Allyn L. Boldt and Scott A. Colby of Westinghouse Hanford Company.

Contributing experts:

Moses Attrep, Jr.	Los Alamos National Laboratory
Steven A. Barker	Westinghouse Hanford Company
Walter D. Bond	Oak Ridge National Laboratory
David O. Campbell	Consultant
Emory D. Collins	Oak Ridge National Laboratory
Harry J. Dewey	Los Alamos National Laboratory
E. Philip Horwitz	Argonne National Laboratory
George Jansen, Jr.	Westinghouse Hanford Company
Gregg J. Lumetta	Pacific Northwest Laboratory
Wallace W. Schulz	Consultant
Major C. Thompson	Westinghouse Savannah River Company
George F. Vandegrift	Argonne National Laboratory

REFERENCES

10 CFR 61, 1988, U.S. Nuclear Regulatory Commission, Licensing Requirements for Land Disposal of Radioactive Waste, "U.S. Code of Federal Regulations."

Beard, S.J., and Caudill, H.L., 1964, "Technetium Recovery and Storage at B-Plant," HW-83348, General Electric Co., Richland, Washington.

Bibler, J.P., and Wallace, R.M., 1987, "Preparation and Properties of a Cesium Specific Resorcinol - Formaldehyde Ion Exchange Resin," DPST-87-647, Savannah River Laboratory, Aiken, South Carolina.

Bibler, N.E., Hoisington, J.E., and Holtzcheiter, E.W., 1981, "Technical Data Summary Decomposition of Oxalic Acid by the Manganese Catalyzed Nitric Acid Reaction," DPSTD-80-36, Savannah River Laboratory, Aiken, South Carolina.

Bond, W. D., 1990, The thorex process, in: "Science and Technology of Tributyl Phosphate," Schulz, W.W., Burger, L.L., and Navratil, J.D., eds., Volume III, CRC Press, Inc., Boca Raton, Florida.

Bray, L.A., Elovich, R.J., and Carson, K.J., 1990, "Cesium Recovery Using Savannah River Laboratory Resorcinol-Formaldehyde Ion Exchange Resin," PNL-7273, Pacific Northwest Laboratory, Richland, Washington.

Bray, L.A., Lust, L.F., Moore, R.L., Roberts, F.P., Smith, F.M., Van Tuyl, H.H., and Wheelwright, E.J., 1964, "Recovery and Purification of Multikilocurie Quantities of Fission Product Strontium by Cation Exchange," Chemical Engineering Progress Symposium Series 47, Volume 60, pp. 9-19.

Buck, J.W., and Peffers, M.S., 1991, "Preliminary Recommendations on the Design of the Characterization Program for the Hanford Site Single-Shell Tanks--A System Analysis," PNL-7573, Vol. 2, Pacific Northwest Laboratory, Richland, Washington.

Buckingham, J.S., et al., 1967, "Waste Management Technical Manual," ISO-100 DEL, Isochem, Inc., Richland, Washington.

Campbell, D.O., and Lee, D.D., 1991, "Treatment Options and Flowsheets for ORNL Low-Level Liquid Waste Supernate," ORNL/TM-11800, Oak Ridge National Laboratory, Oak Ridge, Tennessee.

Delegard, C.H., 1985, "Solubility of PuO_2 x H_2O in Alkaline Hanford High-Level Waste Solution," RHO-RE-SA-75P, Rockwell Hanford Operations, Richland, Washington.

Droppo, J.G., Jr., et al., 1991, "Single-Shell Tank Constituent Rankings for Use in Preparing Waste Characterization Plans," PNL-7572, Pacific Northwest Laboratory, Richland, Washington.

Faubel, W., and Ali, S.A., 1986, Separation of cesium from acid ILW-PUREX solutions by sorption on inorganic ion exchangers, *Radiochimica Acta* 40:49-56.

General Electric, 1964, "Quarterly Progress Report - A Study of Tungsten-Technetium Alloys, April 1, 1964 - July 1, 1964," HW-83550, General Electric Co., Richland, Washington.

Grygiel, M.L., Augustine, C.A., Cahill, M.A., Garfield, J.S., Johnson, M.E., Kupfer, M.J., Meyer, G.A., and Roecker, J.H., 1991, "Tank Waste Disposal Program Redefinition," WHC-EP-0475, Rev. 0, Westinghouse Hanford Company, Richland, Washington.

Horwitz, E.P., 1993, Combining extractant systems for the simultaneous extraction of transuranic elements and selected fission products, in: "Proceedings of the First Hanford Separation Science Workshop, July 23-25, 1991, Richland, Washington, PNL-SA-21775, Pacific Northwest Laboratory, Richland, Washington.

Horwitz, E.P., Dietz, M.L., and Fisher, D.E., 1991, SREX. A new process for the extraction and recovery of strontium from acidic nuclear waste streams, *Solvent Extraction and Ion Exchange* 9:1-25.

Horwitz, E.P., Diamond, H., Gatrone, R.C., Nash, K.L., and Rickert, P.G., 1990, "TUCS: A New Class of Aqueous Complexing Agents for Use in Solvent Extraction Processes," presented at the International Solvent Extraction Conference, ISEC-90, July 20, 1990, Kyoto, Japan.

Horwitz, E.P., Kalina, D.G., Diamond, H., Vandegrift, G.F., and Schulz, W.W., 1985, The TRUEX process - a process for the extraction of the transuranic elements from nitric acid wastes utilizing modified PUREX solvent, *Solvent Extraction and Ion Exchange* 3:75-109.

Jantzen, C.M., 1990, Formation of Zeolite During Caustic Dissolution of Fiberglass: Implications for Studies of the Kaolinite-to-Mullite Transformation, *J. Am. Ceram. Soc.* 73(12):3708-3711.

Koch, G., 1969, "Recovery of By-Product Actinides from Power Reactor Fuels," KFK-976, Gesellschaft fur Kernforschung, Karlsruhe, Germany.

Kolarik, Z., 1991, "Separation of Actinides and Long-Lived Fission Products from High-Level Radioactive Waste (A Review)," KfK 4945, Kernforschungszentrum Karlsruhe, Karlsruhe, Germany.

Lumetta, G.J., and Swanson, J.L., 1993, "Pretreatment of Plutonium Finishing Plant (PFP) Sludge: Report for the Period October 1990-March 1992," PNL-8601, Pacific Northwest Laboratory, Richland, Washington.

Lumetta, G.J., Wagner, M.J., Colton, N.G., and Jones, E.O., 1993, "Underground Storage Tank Integrated Demonstration, Evaluation of Pretreatment Options for Hanford Tank Wastes," PNL-8537, Pacific Northwest Laboratory, Richland, Washington.

Marsh, S.F., and Yarbro, S.J., 1988, "Comparative Evaluation of DHDECMP and CMPO as Extractants for Recovering Actinides from Nitric Acid Waste Streams," LA-11191, Los Alamos National Laboratory, Los Alamos, New Mexico.

Roberts, F.P., Smith, F.M., and Wheelwright, E.J., 1963, "Recovery of Technetium from Hanford Waste," HW-SA-2851, General Electric Co., Richland, Washington.

Schulz, W.W., 1980, "Removal of Radionuclides from Hanford Defense Waste Solutions," RHO-SA-51, Rockwell Hanford Operations, Richland, Washington.

Smith, H.D., Mackey, D.B., Pool, K.H., and Schwenk, E.B., 1992, Corrosion resistance of stainless steels and high Ni-Cr alloys to acid fluoride wastes, in: "Proceedings of the Third International Conference on High Level Radioactive Waste Management," April 12-16, 1992, Las Vegas, Nevada.

Straalsund, J.L., Swanson, J.L., Baker, E.G., Holmes, J.J., Jones, E.O., and Kuhn, W.L., 1992, "Clean Option: An Alternative Strategy for Hanford Tank Waste Remediation, Volume 1. Overview," PNL-8388 Vol. 1, Pacific Northwest Laboratory, Richland, Washington.

Thompson, G.H., Childs, E.L., Kochen, R.L., Schmunk, R.J., Smith, C.M., 1979, "Actinide Recovery from Combustible Waste: The Ce(IV)-HNO_3 System Final Report," RFP-2907, Rocky Flats Plant, Golden, Colorado.

U.S. Department of Energy (DOE), 1987, "Final Environmental Impact Statement - Disposal of Hanford Defense High-Level, Transuranic, and Tank Wastes," DOE/EIS-0113, Richland, Washington.

Vialard, E., and Germain, M., 1986, "Technetium Behavior Control in the Purex Process," ISEC '86 International Solvent Extraction Conference Preprints, p. I-137, DECHEMA, Frankfurt am Main, Germany.

Wheelwright, E.J., Bray, L.A., Van Tuyl, H.H., and Fullam, H.T., 1974, "Flowsheet for Recovery of Curium from Power Reactor Fuel Reprocessing Waste," BNWL-1831, Pacific Northwest Laboratory, Richland, Washington.

INDEX

Bis-4,4'(5')-t-butyldicyclohexano 18-crown-6, 81, 97, 184, 190-193

Clean Option, 155
  acid-side separations, 173-187
    Cs removal using AMP in, 187
    solvent extraction cycle using CMPO in, 180
    solvent extraction cycle using crown ether in, 184
    solvent extraction cycle using TBP in, 176-180
    treatment of low-level waste streams from, 193-195
  base-side separations, 187
    Cs removal in, 190-191
    NRC limits of wastes, 161
    preparation of feed for, 187-189
    Sr removal in, 190-193
    Tc removal in, 193
    treatment of low-level waste streams from, 195-199
  objectives, 157
    disposal of HLW, 157
  separation requirements, 12, 13, 160-199
    flowsheets, 165
    sludge dissolution, 169
    sludge washing, 169
  tank waste, 158-159
    composition of, 159
    disposition of constituents, 165, 166
    important radionuclides in, 164
    types of, 158
  vitrification plant, 160
    allowable additions to glass for, 161
    canister generation from, 165

Clostridium SP, 117-120, 130

Contaminated soils, 115
  treatment of
    microbiological processes, 115-130
    SOIL*EX^{SM} process, 145-154

Formic acid, 101-113
  application in waste vitrification systems, 101
  simulated Savannah River Site sludges, 104

Formic acid (cont'd)
  catalytic decomposition of, 102
    effects of nitrite ion on, 106-108
    experimental studies of, 103, 104
    hydrogen gas evolution, 101, 102, 106-108
    noble metals as catalysts, 102, 103, 105-108
    reaction mechanisms for, 110-112

Hanford Site
  double-shell tank sludges
    acid dissolution of, 39, 41
    application of TRUEX process to, 41-44, 48
  double-shell tank wastes
    complexant concentrate, 27
    composition of, 27
    description of, 25, 26
    inventory, 25, 27
    neutralized cladding removal waste (NCRW/, 27
    neutralized current acid waste (NCAW), 27
    plutonium finishing plant (PFP waste), 27
    pretreatment for disposal, 33-35
    strategy for disposal of, 28-31
  history of, 5
  location of, 25
  single-shell tank sludges
    selective leaching of chromium, 38-41, 43,44, 48
    selective leaching of aluminum, 40, 41, 48
  single-shell tank wastes
    composition of, 7, 8, 27, 28
    description of, 6, 25, 26
    inventory, 25, 27
    pretreatment for disposal, 9, 10, 33-35
    selective leaching of phosphate, 40, 41, 48
    selective leaching of sulfate, 40, 41, 48
    strategy for disposal of, 11, 12, 28-31
  tank waste remediation system, 12, 31, 32
  tri party agreement, 11, 12, 29

Hanford Tank 241-SY-101, 71-79
  composition of waste in, 72, 73
    inorganic components, 72
    total organic carbon, 73
  episodic release of flammable gas, 72-78
    gas composition, 72, 75
    radiolysis effects, 76, 77
    reaction mechanism for thermal effects, 73-78
  history of, 71

Idaho Site, 5
  calcined waste, 5
    future disposal of, 5
    inventory of, 5
    origin of, 5

Integrated fast reactor
  description of, 133
  electrorefining of irradiated fuel, 133-139
    chemical basis for, 135-138
    description of, 133, 134
    forms for salt waste, 141
    treatment of cadmium fraction, 135, 139, 141
    treatment of product salt, 133-135
    zeolite treatment of stripped salt, 139-141
  fuel for, 133
  pyrochemical fuel reprocessing, 133
    development status, 141, 142

Microbiological remediation technology, 115-132
  action on soils and sediments, 115-132
  anaerobic bacterial processes, 116
    action of *clostridium sp*, 117-120, 130
    action of *pseudomonas fluorescens*, 121-123, 127, 129
    dissolution of metals, 119-120
    precipitation of metals, 119-120
  citric acid processes
    biodegradation of sludge extracts, 126-129
    extraction of metals, 126
    photodegradation of citrates, 127-129

Nuclear Waste Minimization, 13-14
  philosophy for, 13
  separation technology, key to, 1-3, 13-14

Oak Ridge National Laboratory Site, 4, 6, 8-12
  storage of nuclear waste, 4, 6, 8-11
    composition of, 4, 6, 8
    future disposal of, 8, 11
    gunnite tanks for disposal of, 4
    hydrofracture disposal of, 4, 5
    inventory of, 4, 8

Octyl(phenyl)-N,N-diisobutylcarbamoylmethyl-
    phosphine oxide (CMPO), 40-44, 48, 61-64, 55-59, 81-98, 180

Pseudomonas fluorescens, 121-123, 127-129

Removal of actinides, 51
  column tests, 61-68
    capacity for Am, 61, 62
    extraction and elution kinetics, 62-64
    on actual waste, 64-68
  distribution coefficients, 56
    batch measurements of, 55
    column measurements of, 55
    effects of U and Bi on Am, 58-59
  extraction chromatography, 53
    comparison of materials for, 59
    large-scale processing using, 68-69
    materials for, 53

Removal of actinides (cont'd)
  Hanford waste, 54
    chemical components in, 54
    radionuclides in, 54, 55

Savannah River Site
  high-level waste, 17-21
    composition of, 17
    dissolution of aluminum, 19
    inventory of, 17
    pretreatment of, 17-19
  in-tank precipitation process, 17-20
    description of, 20
    process chemistry, 20, 21
  location of, 17

Sodium titanate, 21
  filtration of, 22
  precipitation of, 21
  sorption of actinides by, 21
  sorption of ^{90}Sr by, 21

SOIL*EX^{SM} Process, 145-154
  application to soils, 145, 154
    candidate DOE site soils, 154
  components of
    ACT*DE*CON System technology, 145, 146, 148
    dewatering subsystem, 149, 150
    extraction subsystem, 147-149
    PO*WW*ER System technology, 150, 151
  description of, 145
development status of, 151-153

Solvent extraction processes
  TRUEX-SREX extraction/recovery process, 81
    acid concentration profile, 96
    Am concentration profiles in, 97
    flowsheet development for, 95-98
    flowsheet testing of, 96
    Pu concentration profiles in, 98
    Sr concentration profiles in, 97
    Tc concentration profiles in, 98
  combined process solvent, 86
    chemical properties of the, 91-94
    physical properties of the, 86-90
  selective partitioning, 94
    tetrahydrofuran tetracarboxylic acid used for, 94

Tetraphenylborate salts, 20-22
  chemical stability of, 21, 22
  filtration of, 22
  flammability hazards of, 22
  precipitation from wastes, 20
  radiation stability of, 21, 22

TRUEX process
  application to NCRW sludge, 44-48
  chemical flowsheet for, 42
  organic solvent composition, 40, 41

212